Tae Hyun Gil

Development of alternative processes for top electrode deposition on OLED based on physical vapour deposition process

AF875391

TUD*press*

Tae Hyun Gil

Development of alternative processes for top electrode deposition on OLED based on physical vapour deposition process

TUD*press*
2011

Bibliografische Information der Deutschen Nationalbibliothek
Die Deutsche Nationalbibliothek verzeichnet diese Publikation in der Deutschen Nationalbibliografie; detaillierte bibliografische Daten sind im Internet über http://dnb.d-nb.de abrufbar.

Bibliographic information published by the Deutsche Nationalbibliothek
The Deutsche Nationalbibliothek lists this publication in the Deutsche Nationalbibliografie; detailed bibliographic data are available in the Internet at http://dnb.d-nb.de.

ISBN 978-3-942710-54-1
© 2011 TUDpress
Verlag der Wissenschaften GmbH
Bergstr. 70 | D-01069 Dresden
Tel.: 0351/47 96 97 20 | Fax: 0351/47 96 08 19
http://www.tudpress.de

Alle Rechte vorbehalten. All rights reserved.
Gesetzt vom Autor.
Printed in Germany.

Technische Universität Dresden

Development of alternative processes for top electrode deposition on OLED based on physical vapour deposition process

Tae Hyun Gil

der Fakultät Elektrotechnik und Informationstechnik der Technischen Universität Dresden

zur Erlangung des akademischen Grades eines

Doktoringenieur

(Dr.-Ing.)

Vorgelegte Dissertation

Verteidigt am : 5. 4. 2011

1. Gutachter: Prof. Dr. Hubert Lakner

2. Gutachter: Prof. Dr. Karl Leo

Contents

1 Introduction

Electroluminescence from organic materials was produced first by A. Bernanose and coworkers at the Université de Nancy, France in the early 1950s [1]. Since Tang and Slyke demonstrated a multilayered organic electroluminescence device [2], OLEDs have attracted great attention and achieved innovative developments in the last two decades. OLEDs are applicable to various optoelectronic devices due to their self-emissive characteristics from thin organic layers. They also have natural color expression, and a simple device structure with a thin device thickness of less than 3 mm. The initial products of OLEDs were sub-displays in cellular phones and front panels of car audios, which were relatively small information devices. A large OLED display with a homogeneous emission from each pixel can be realized combined with a thin film transistor (TFT) backplane technique similar to liquid crystal displays (LCDs), and has already been successfully commercialized to the cellular phone market as main display. OLED must compete with the LCDs for price, size and reliability. However, OLEDs have many advantages compared with TFT-LCD, for example, 1,000 times faster response time, low power consumption, simple structure and processes. In spite of those advantages, organic materials used in OLED fabrication are still expensive and have a lower physical and chemical stability than inorganic semiconductors. However, an ultra long-life OLED was also reported using stable host materials for a phosphorescent emitter material and doped transport layers [3], and it is believed that further improvement is possible for long-living OLEDs.

Large area general lighting is also an attractive field that OLEDs can challenge current fabrication technologies. The most intensive research for OLED lighting is being performed in the European Union in order to replace incandescence bulb- and lamp-based current lighting in the near future. As a representative case, an integrated project by the European Commission "High brightness OLEDs for ICT & Next Generation Lighting Applications" is being performed in industry and institutes for the development of highly efficient and large area white OLEDs. According to a recent report, the OLED lighting market will start in 2011. Revenues with OLED lighting are forecasted to surpass passive matrix OLED displays in the 2013/2014 time frame, reaching $6 billion by 2018 [4].

OLEDs basically consist of several functional organic layers, e.g. transport layers, emission layers, etc., between two electrodes. Organic thin films and top electrodes are usually deposited by vacuum evaporation on glass substrates, where a transparent electrode is previously deposited. If large area substrates and deposition equipment are provided, they enable a large emission surface depending on the substrate size. Not only material developments, but the developments of deposition techniques are also a critical issue that must be improved for mass-production. Important points in film deposition are homogeneous thickness and reproducibility. The thickness of organic layers affects the carrier transport in organic layers, which can result in differences in the driving voltage and the quantum efficiency. Since the OLED is basically a current-driven device, the conductivity of the electrodes plays a significant role especially in large area OLED. Like the effect of the organic layers, an uneven electrode thickness can also yield a different charge transport. If a large active area is realized as a single segment, a difference in luminance can occur, depending on the conductivity and thickness of the electrode.

Organic materials are evaporated mostly at temperatures between 150 to 500°C. In contrast, metals used in OLED need much higher temperatures than organic materials. For instance, Al and Ag are vaporized, respectively, above 820 and 950°C at a given vapor pressure of 1×10^{-6} Torr. There are various kinds of materials and designs for heating elements, but heater reliability and process stability are still problematic because such heating elements suffer from an extremely high electrical power density in order to reach a feasible temperature. This leads to a fast material degradation, even worse for larger elements, and results in an increase of equipment downtime due to frequent maintenance. Although many types of evaporators were applied in OLED mass production, alternative electrode deposition techniques on a substrate of size larger than 2nd generation (370 × 470 mm^2) are still strongly demanded for the production of large area OLEDs with long equipment uptime as well as good film homogeneity.

The main focus of this study is to develop alternative processes for the top electrode deposition on OLED by means of a magnetron sputter method. Magnetron sputtering has been a strong candidate for top electrode deposition because it is a very versatile physical vapor deposition (PVD) technique. In principle, any solid metal or alloy and a variety of compounds can be deposited on various types of substrate by the magnetron sputter technique. The prominent advantages are the easy enlargement of sputtering cathode size for large area depositions and the long process stability. These led to the first motivation of this study to apply the magnetron sputter technique for the deposition of metal electrode on organic layers.

Also, in a reactive sputtering, the stoichiometry of a compound film is easily optimized by controlling the mixing ratio of an inert gas and a reactive gas. The reactive sputtering has already been used in OLED production for the deposition of transparent conducting oxides (TCOs) as the transparent electrode on glass substrate. If the TCO layer is deposited on organic layers, a transparent or top-emission OLED can be realized and applied to transparent information devices in aircraft or automobiles. This is the second motivation of this study.

Some works have used the magnetron sputtering for the deposition of top electrodes. However, it was reported that the OLEDs with the sputtered electrode exhibited serious damage such as high leakage currents and short lifetimes. The reasons responsible for the damage were explained with molecular defects induced by atomic energetic particles or secondary electrons generated during sputtering process [5, 6]. However, those particles have to be taken into account together in order to distinguish corresponding damage sources because they are generated simultaneously.

The first part (section 6.1.1) of this study, therefore, contributes to the identification of degradation mechanisms using conventional magnetron sputtering and various characterization methods to explain the results. For the experiments, Al films were deposited on organic layers with various process parameters, and the characteristics of OLEDs were observed in detail.

In order to reduce the damage, various power sources and sputtering types were employed (section 6.1.2, 6.1.3, and 6.1.4). The first attempt was to use radio frequency (RF) super-imposed DC power. By the addition of the RF to the DC power, the discharge voltage decreases as the portion of the RF power increases. This effect can reduce the kinetic energy of the particles, but the leakage currents were not significantly decreased in spite of the decrease of the discharge voltage.

A special sputtering system, so-called facing target sputter (FTS), was used. This technique was reported by other researchers to be helpful for organic layers to minimize the damage from the bombardment of the energetic particles [7-10]. In this study, a linear type FTS was designed to cover the substrate size of 200×200 mm^2. An Al target was used to deposit Al top electrode. Although FTS was expected to realize a damage-free sputtering method, the results showed that the degradation in OLEDs can not be completely eliminated with FTS as well.

However, a low damage sputtering was accomplished by means of a dual magnetron sputter with bipolar-pulsed DC and hardware modifications in the Inline equipment. This technique is expected to be applicable for the deposition of the top metal electrode in the OLED mass-production.

FTS was also used for TCO deposition on organic layers (section 6.2). Indium tin oxide (ITO) was used as the target material, and the same process parameters were applied to compare the properties of the ITO films with those of the Al films. After the film characterization, the ITO films were deposited on organic layers with and without a thin Ag layer, which can act as a primary electrode and a buffer layer against the sputter damage. Various experiments were performed to realize a low-damage ITO deposition, but the results were unfortunately not fully satisfying. Although the low-damage ITO sputtering was not completely successful, a possible method for the realization of highly transparent cathode on organic layers is suggested on the basis of the result.

2 Physics of OLED

This section deals with the general information on OLEDs. In the first subsection, electronic and optical properties of organic materials are introduced, focusing on the small molecule organic materials typically used in OLEDs, and the mechanisms of electroluminescence from excited organic molecules are also described. In the next subsection, OLED structures and typical characteristics are discussed together with the process techniques for the materials used for OLEDs.

2.1 Organic semiconductor

2.1.1 Basic properties of organic semiconductor

Organic materials in electronics are classified into two groups according to their molecular weights: small molecular organic materials and polymers. Many organic compounds contain delocalized electrons provided by a conjugated π-electron system. As shown in Figure 2.1, benzene has six sp^2 hybridized carbon atoms having bond angles of 120°. Each of its carbons uses two sp^2 orbitals to bind to two other carbons. Because benzene is planar, the six p_z orbitals are parallel and close enough to overlap by each other. As a result, the overlapping p_z orbitals form a continuous cloud of π electrons above and below the plane of the benzene ring. Each of the six π electrons is localized neither on a single carbon nor in a bond between two carbons. Instead, each π electron is shared by all six carbons. Therefore, the six π electrons are delocalized.

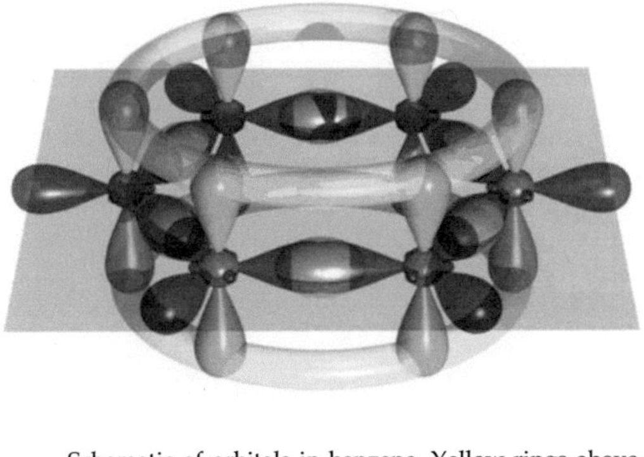

Figure 2.1: Schematic of orbitals in benzene. Yellow rings above and below the plane are π electron system formed by overlapping of six p orbitals (sky blue lobes). The carbon-carbon and the carbon-hydrogen bonds are σ bonds. Reprinted from Ref. [11].

Since the overlapping of the p carbon orbitals is smaller than that of the sp^2 carbon orbitals, the resulting binding π and anti-binding π^* benzene orbitals split up less than σ and σ^* bonds. Therefore, the π binding orbital is the highest existing state, and the lowest energy transitions take place in the π-π^* orbitals. Normally, the π orbital is termed the highest occupied molecular orbital (HOMO) and the π^* orbital is the lowest unoccupied molecular orbital (LUMO). The HOMO is also denoted as the ionization energy of the molecule and the energetically higher π^* orbital corresponds to the electron affinity. The energy gap between HOMO/LUMO ranges between 1.5 and 3 eV, leading to light absorption or emission in the visible spectral range [11]. This energy gap is usually called band-gap, and it influences the electrical and optical properties of the molecules. As the number of the benzene ring increases in a polycyclic molecule, the π electron system becomes larger. It results in an increase of the delocalization of the electrons and a decrease of the band-gap.

The organic semiconductors used in OLEDs are usually grown as a disordered amorphous phase so that the band structures like in inorganic crystalline semiconductors are not formed. The optical absorption and emission properties are therefore determined by the physical properties of the isolated molecules. In order to excite a molecule, a photonic energy corresponding to the band gap of the molecule or larger is required. When the photonic energy is transferred, an electron is promoted from the π orbital to the energetically higher π^* orbital, leaving a hole behind. Due to strong Coulomb interaction between electron and hole, they form an electron-hole pair, a so-called exciton. Since the delocalization of electrons in organic molecules is much weaker than in inorganic materials due to weak intermolecular van der Waals interaction, the excitons are typically localized on individual molecules. As a result, the exciton has a high binding energy of up to 1 eV, and the optical properties are dominated by transitions of Frenkel excitons. In contrast, the excitons in inorganic solids have binding energy of only few meV, and these excitons are called Wannier-Mott excitons which are more delocalized in a solid. The two kinds of exciton are illustrated in Figure 2.2.

The Wannier-Mott exciton has a large exciton radius ranging from 4 to 10 nm [13], while the Frenkel exciton is almost localized to a molecule with a radius of typically under 1 nm [14], which corresponds to the molecular size. Frenkel exciton on an organic molecule is simply an excited molecule. According to the Pauli principle, the electrons possess half-integer spin $S = 1/2$, and no two electrons can occupy a same quantum state with a same spin quantum number. If the total spin quantum number of two unpaired electrons is $S = 0$ in an excited state, the total wave function is antisymmetric for two fermions. The molecule is then in a singlet state S, as indicated in Figure 2.2 (b).

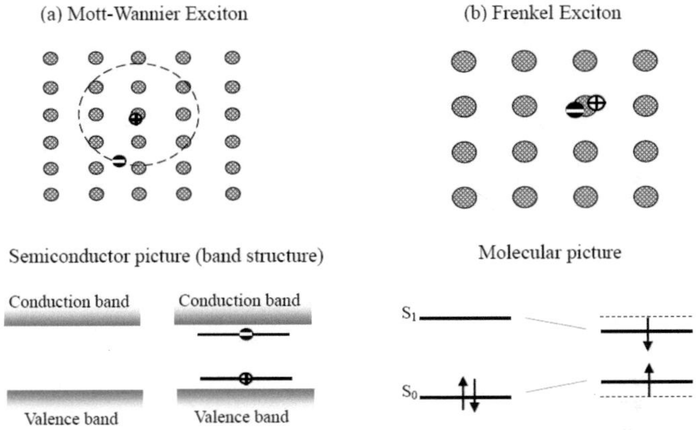

Figure 2.2: Schematic illustration of Wannier-Mott exciton with the band diagram of an inorganic semiconductor (a), and Frenkel exciton with the energy level diagram of individual molecular states (b).

The singlet ground state $^1\Phi_0$ and the excited singlet state $^1\Phi_1$ is written as products of a symmetric orbital function and an antisymmetric spin function:

$$^1\Phi_0 = \phi_1(1)\phi_1(2) \cdot \sqrt{\frac{1}{2}}[\alpha(1)\beta(2) - \beta(1)\alpha(2)] \tag{2. 1. 1}$$

$$^1\Phi_1 = \sqrt{\frac{1}{2}}[\phi_1(1)\phi_2(2) + \phi_2(1)\phi_1(2)] \cdot \sqrt{\frac{1}{2}}[\alpha(1)\beta(2) - \beta(1)\alpha(2)] \tag{2. 1. 2}$$

where ϕ_n is the orbital of n-electron, and α and β denote the orthonormal spin functions.

If the spin is symmetric, the total spin quantum number is S = 1. This state is called triplet state, denoted as T. It can have three possible spin states. The excited triplet state wave functions can be written as products of an antisymmetric orbital function and a symmetric spin function:

$$^3\Phi_1 = \sqrt{\frac{1}{2}}[\phi_1(1)\phi_2(2) - \phi_2(1)\phi_1(2)] \cdot \begin{cases} \alpha(1)\beta(2) & , m_s = 1 \\ \sqrt{\frac{1}{2}}[\alpha(1)\beta(2) + \beta(1)\alpha(2)] & , m_s = 0 \\ \beta(1)\alpha(2) & , m_s = -1 \end{cases} \tag{2. 1. 3}$$

The organic molecules in a film form can be also electrically excited, creating Frenkel excitons. If an electric field is applied through the film which has contacts at both sides, charge carriers are injected from the contacts. Additional electrons and holes move respectively on the LUMO and HOMO in opposite direction. Thereby, a screening of the additional charge occurs through corresponding alignment of the charge densities of the neighboring molecules. It leads to an electronic polarization of the molecules, and the charge together with the induced polarization is called polaron, which forms polaron states. The polarization is a very fast process with a relaxation time between $10^{-16} \sim 10^{-15}$ s. It is a much shorter time compared to the typical moving time of charge carriers, and is thus the dominant interaction of charge carriers with molecular lattices, having an energy of typically 1.5 ~ 2.0 eV [15]. Due to the polarization energy in a solid, the transport levels become energetically lowered, and the exciton has further lower energy because of the binding energy. Therefore, the HOMO and LUMO levels denote charge transport levels in the solid, and the bound singlet and triplet states denote the exciton levels, as illustrated in Figure 2.3.

The energy levels and states discussed above are illustrated in Figure 2.4, the so-called Jablonski diagram. The Jablonski diagram depicts the electronic states which involve absorption, radiative, and nonradiative transitions. The radiative transitions are indicated by straight arrows, and the nonradiative transitions are by squiggly arrows. The radiative energy transitions in organic molecules obey Kasha's rule that the energy transitions occur from the lowest excited states (S_1 or T_1) to the ground state (S_0) [16]. The radiative transition from S_1 to S_0 is termed 'fluorescence', and from T_1 to S_0 is termed 'phosphorescence'. They will be described in detail in next subsection. The non-radiative transitions occur when the electron is excited to higher states S_{n+1}, where $n \geq 1$, than S_1. Within the same spin multiplicity, electrons relax the energy from the S_{n+1} to S_1 through very fast internal conversion (typically 10^{-13} s), losing energy via vibrational relaxation at the same time scale as molecular vibrations. Energy transfer between the states of different multiplicities via spin-inversion is called intersystem crossing, and this energy transfer from S_1 to T_1 is very efficient.

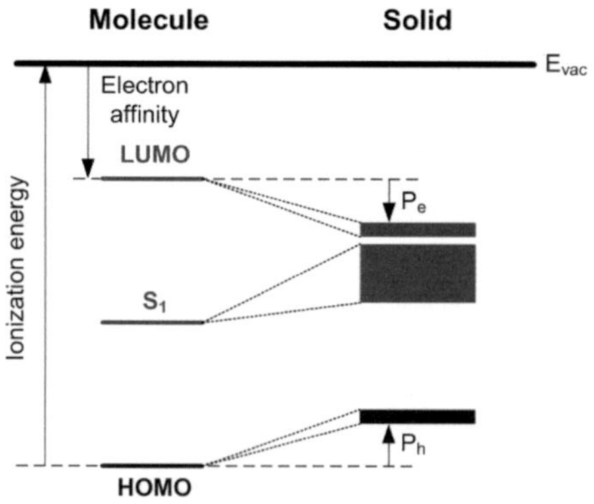

Figure 2.3: HOMO and LUMO energy levels in a molecular and solid state. By the polarization energy, P_e and P_h for electron and hole, the transport levels of charge carriers are lowered. The exciton levels become lowered further due to the exciton binding energy. Reprinted from Ref. [17].

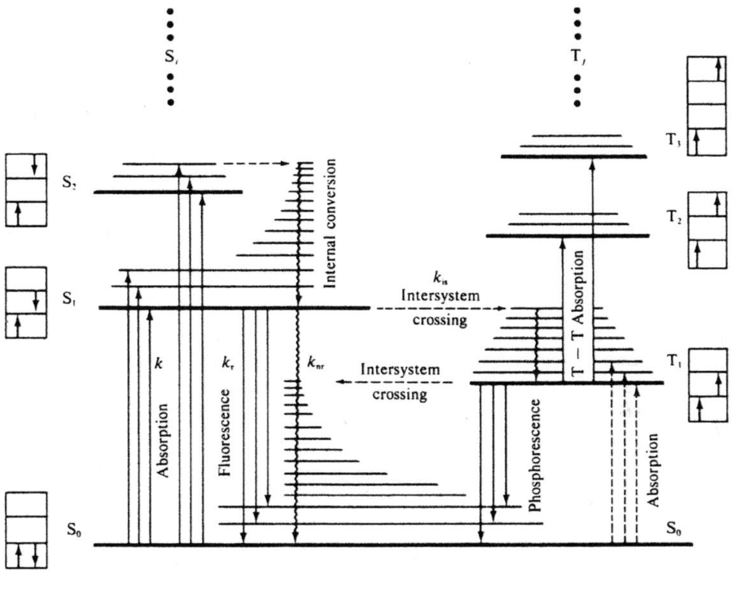

Figure 2.4: Jablonski diagram of an organic molecule. The diagram illustrates absorption, radiative and non-radiative energy transition processes. Reprinted from Ref. [18].

2.1.2 Electroluminescence

Electroluminescence is a phenomenon in which an electrically excited material emits light in response to an electrical current or an electric field. In organic materials, Frenkel excitons are formed spin-independently by injected electrons and holes from electrodes. The ratio of the singlet to triplet excitons is determined by spin statistics. 25 % are singlets, and 75 % are triplets [19]. For small organic molecules, this 1:3 ratio has been also accepted experimentally [20]. This singlet ratio results in a limitation of the external quantum efficiency of around 5 % when a fluorescent emitter is used. In order to improve the efficiency of OLED, both singlet and triplet excited states are required to contribute to luminescence, because triplet states constitute the majority of electrogenerated excited states. However, the probability of luminescence from the triplet state is in general very low due to spin-forbidden nature such that almost all its energy is lost to non-radiative processes. Although radiation from the triplet state is rare, radiative process can be enabled by incorporating a heavy metal atom, such as Pt and In, into organic complexes [19]. In metal-organic complexes, a strong spin-orbit coupling leads to singlet-triplet states mixing, so highly efficient phosphorescence is achieved with internal efficiency of nearly 100 %.

As a method to achieve high efficiency and charge transport of an emission layer in OLED, a small concentration of a phosphorescent emitter as a 'guest' is mixed into a 'host' material having suitably high conductivity. This method is advantageous to confine excitons to the phosphorescent emitters so that exciton quenching by other excitons can be minimized [21].

There are two mechanisms for energy transfer between molecules: Förster and Dexter energy transfer. Förster energy transfer is the mechanism to transfer energy of a donor (D) to an acceptor (A) through resonant dipole-dipole transition, which is a long-range (4 ~ 10 nm) energy transfer [22]. The transfer rate efficiency is influenced by the distance between the donor and acceptor, the spectral overlap of the donor emission and the acceptor absorption, and the orientation of the donor emission dipole moment and the acceptor absorption dipole moment. In Förster transfer, spin is not changed, thus

$^{1}D^{*} + {}^{1}A \rightarrow {}^{1}D + {}^{1}A^{*}$

$^{1}D^{*} + {}^{3}A \rightarrow {}^{1}D + {}^{3}A^{*}$

are allowed. The superscript 1 and 3 represents singlet and triplet state, respectively, and the asterisk signifies an excited state. The triplet state of the donor can also Förster transfer to the singlet state of the acceptor, expressed by

$^{3}D^{*} + {}^{1}A \rightarrow {}^{1}D + {}^{1}A^{*}$

This process is known as a phosphor sensitized fluorescence, and it results in the transfer of triplet excitons of donor to the singlet state of a fluorescent dye [23].

Dexter energy transfer occurs at short distances between donor (D) and acceptor (A) molecules such that electron exchange interaction takes place at a rate proportional to the spatial overlap of the orbitals of D and A molecules. The total spin from matrix and emitter molecule is conserved, and two transitions are allowed:

$^{1}D^{*} + {}^{1}A \rightarrow {}^{1}D + {}^{1}A^{*}$

$^{3}D^{*} + {}^{1}A \rightarrow {}^{1}D + {}^{3}A^{*}$

namely, singlet-singlet and triplet-triplet energy transfer, respectively. Since Förster energy transfer dominates singlet-singlet transition for larger distances, Dexter singlet-singlet transitions are usually not observed due to the short distance energy transfer (~ 1nm) by the close distance between molecules. Therefore, Dexter energy transfer is the important mechanism for energy transfer between triplets.

Light is generated by the radiative decay of excitons in molecules. OLEDs use various kinds of organic material including both fluorescent and phosphorescent emitters in order to achieve various colors and high efficiency in visible range. Radiation is produced by the energy loss of excitons. This is achieved by Langevin type recombination regardless of the spin orientation [24]. The recombination rate is given by

$$\kappa_{np} = \frac{e}{\varepsilon\varepsilon_0}(\mu_n + \mu_p) \qquad (2.1.4)$$

where e is the elementary charge, $\varepsilon\varepsilon_0$ is the permittivity of the organic layer, and μ_n and μ_p are the electron and hole mobility, respectively. This rate is dependent on whether the organic material is electron-transporting or hole-transporting, and also on the injected electron and hole currents. The recombination efficiency is therefore influenced by the injected charge carrier balance, and can be ideally 1, if electrons and holes are injected equally. This is one of the factors to determine the external efficiency of OLED, and will be discussed in section 2.2.3.

The radiation lifetime of singlet excitons for a fluorescent transition is typically on the order of 10^{-9} s. On the other hand, the lifetime of triplet excitons for a spin-forbidden phosphorescent transition can be as long as several seconds. As discussed above, a radiative transition is possible by enhancing spin-orbit coupling, shortening the lifetime to 10^{-6} s.

2.1.3 Charge transport in organic semiconductors

Both driving voltage and luminance efficiency of OLEDs are strongly affected by charge injection from the electrodes and charge transport in the organic materials. To achieve a good injection, it is beneficial to have an ohmic contact between the organic layer and the charge-injecting contact, and the ohmic contact can be realized by a doped organic layer [25]. For an intrinsic semiconductor, a Schottky contact is formed at the interface such that electrode materials with low work function, for example, Yb or Ca, are widely used to improve the injection into the organic layers [26, 27].

After the charge injection, the charge transport in organic layers plays an important role in optimizing the device efficiency of OLED. The organic materials used for OLEDs are generally amorphous after evaporation and, thus, treated as a disordered film. Unlike highly ordered semiconductors, in which charge carriers transport over valance and conduction bands with high carrier mobility, the amorphous organic materials have no well-defined band structures so that the charge transports via hopping between localized states [28], resulting in low carrier mobility. The carrier mobility strongly depends on temperature, electric field, and charge carrier density [29, 30]. The electric field dependent mobility follows a Poole-Frenkel formula:

$$\mu(E,T) = \mu_0 \exp(\beta\sqrt{E}) \quad (2.1.5)$$

where μ_0 is the zero-field mobility, β is the field enhancement factor, and E is the electric field. The intrinsic carrier density in a semiconductor with an energy gap E_g and an effective density of states N_0 is given by

$$n_i = N_0 \exp(-E_g / 2kT) \quad (2.1.6)$$

For typical values of E_g=2.5 eV and N_0=10^{21} cm^{-3} in an organic semiconductor, a carrier density of n_i=1 cm^{-3} at room temperature is obtained [31]. Furthermore, organic materials have low mobility (10^{-8} to 10^{-2} $cm^2/V\cdot s$) because charge carriers are localized in the molecules, and electron mobility in organic materials is even two or three orders of magnitude lower than hole mobility [32, 33]. The limiting factors that affect the driving voltage and efficiency of OLED are in general the injection and the transport of electrons from the cathode to the emission zone.

Charge carriers are not only localized, but also trapped in local states generated by impurities or film defects. These trap states provide a higher energy tail in the Gaussian distribution of HOMO level acting as a shallow trap, which has energy in the range of 110~150 meV [34]. The charge carriers injected from electrodes at low voltage can be trapped in those defects, deep levels at the contacts, or at hetero junctions. The traps are gradually filled with the increase of electric field. The current begins to increase faster than quadratic with the electric field until all traps are filled. The traps have an exponential distribution, and the trap density of states is described by

$$n_t(E) = \left(\frac{N_t}{kT_t}\right)\exp\left(\frac{E-E_c}{kT_t}\right), \qquad E_c > E > 0 \tag{2. 1. 7}$$

where E_c is the energy of conduction band, N_t the total density of traps, and kT_t an energy characterizing the trap distribution. N_t for *tris*(8-hydroxyquinoline)(Alq_3) based device was estimated to be approximately 10^{20} cm^{-3} [35]. The current in this case is a trap-charge limited current (TCLC), which is given by

$$J_{TCLC} = N_C \mu q \left(\frac{\varepsilon_r \varepsilon_0}{qN_t}\right)^r \left(\frac{2r+1}{r+1}\right)^{l+1} \frac{V^{r+1}}{L^{2r+1}} \tag{2. 1. 8}$$

where $r=T_t/T$, $\varepsilon_r\varepsilon_0$ is the permittivity of organic layer, and the effective density of states in the conduction band $N_C=2.5 \times 10^{19}$ cm^{-3}. When the traps are filled with charge carriers, the organic materials behave as a trap-free ideal space charge limited insulator. The current flowing in OLED is called space charge limited current (SCLC). The current (I)-voltage (V) relation is described by the Mott-Gurney law [36] as

$$J_{SCLC} = \frac{9}{8}\varepsilon_r \varepsilon_0 \mu \frac{V^2}{d^3} \tag{2. 1. 9}$$

where d is the thickness of the device. The current density accounting for the Poole-Frenkel mobility is given as

$$J^{PF} = \frac{9}{8}\varepsilon_r \varepsilon_0 \mu_0 \frac{V^2}{d^3}\exp\left(0.89\beta\sqrt{\frac{V}{d}}\right) \tag{2. 1. 9}$$

This expression is, however, relevant only for a single charge carrier transport and simple heterojunction devices. The state-of-the art OLEDs yield more complicated situations because of a multilayered structure, bipolar charge carriers, exciton formation and recombination, etc. The continuity equation in a one-dimensional device becomes

$$\frac{dJ_e}{dx} = -\frac{dJ_h}{dx} = q\kappa_{np} np \tag{2. 1. 10}$$

where J_e and J_h are electron and hole current density, respectively, κ_{np} the recombination rate, n and p are electron and hole density, respectively. The solutions as a function of voltage are therefore only possible where the mobility is independent of the electric field and the absence of diffusion and trapping. However, some organic materials like Alq_3 show strong field-dependent mobility [37], so the exact analysis is very difficult due to the complexity of the numeric procedures.

2.2 OLED characteristics

2.2.1 OLED structures and materials

2.2.1.1 OLED structures

OLEDs basically consist of organic layers and two electrodes. Since the first OLED was demonstrated by C. W. Tang and S. A. VanSlyke with double organic layers [2], OLEDs have been tremendously improved in efficiency by implementing multilayer structures, exciton blocking layers, and doped transport layers, etc. [38-40].

In a two-layer OLED, the organic layers are chosen to transport holes and electrons, respectively. Organic layers which transport primarily holes and electrons are termed hole transport layer (HTL) and electron transport layer (ETL), respectively. There are several materials that have been preferred as HTLs. *N,N'*-diphenyl-*N,N'*-bis(3-methylphenyl)(1,1'-biphenyl)-4,4'-diamine (TPD) and *N,N'*-bis(naphthyl)-*N,N'*-diphenyl-1,1'-biphenyl-4,4'-diamine (NPB) have been studied extensively. For ETLs, *tris*(8-hydroxyquinolinato)aluminum (Alq_3) has been used widely due to its high electron mobility with a value of approximately 10^{-6} $cm^2/V{\cdot}s$ at $4{\times}10^5$ V/cm, green emission having a peak at 530 nm, and high thermal stability [41].

An energy level diagram and a schematic diagram of the two-layer OLED structure are shown in Figure 2.5. This OLED structure is called bottom emission OLED (BEOLED), in which a generated light passes through the bottom electrode and the glass substrate. In order to pass the generated light through the glass substrate, the bottom electrode must be transparent in the visible range. As the bottom electrode on glass substrates, transparent conducting oxides (TCOs), e.g. indium tin oxide (ITO), indium zinc oxide (IZO), or aluminum zinc oxide (AZO), are widely used [42, 43]. For top electrodes, highly reflective and conductive metals should be chosen. When a heterojunction is formed, facilitating hole-injection from the HTL into the ETL and blocking electron injection in the opposite direction, the probability of exciton formation can be enhanced. Recombination can occur primarily near the interface region so that a green electroluminescence is generated. This is possible because the HOMO (5.5 eV) of the TPD is slightly above that (5.7 eV) of the Alq_3 such that holes can easily enter into the ETL. On the other hand, the LUMO (3.0 eV) of the Alq_3 is significantly below that (2.3 eV) of the TPD so that electrons can be confined in the ETL [44, 45]. By placing the recombination zone at a sufficient distance from the contact, the probability of quenching near the metallic surface can be effectively reduced.

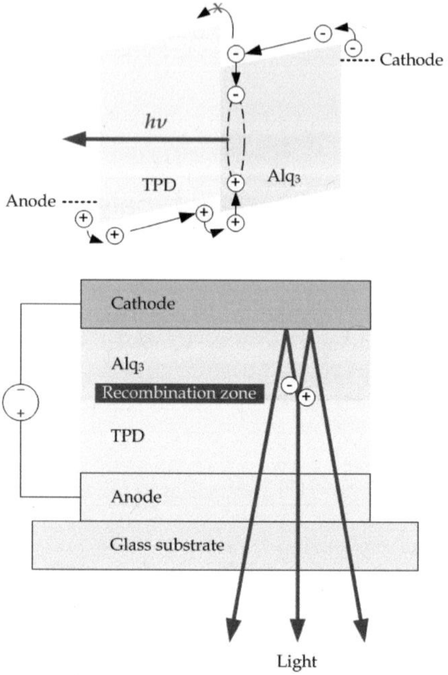

Figure 2.5: Energy band diagram (top) and the structure of two-layer bottom-emission OLED (BEOLED).

The efficiency of the simple OLED can be improved further by employing a three or more layer structure. Thin (~10 nm) organic layers can be introduced between the transport layers and the emission layer to block the excitons within the emission layer [39]. This method is especially helpful for OLED where phosphorescent emitters are used. Since the diffusion lengths of singlet excitons are much shorter (tens to hundreds Å) than those (>1000 Å) of triplet excitons [19, 46], it is essential to confine the triplet excitons within the emission layer. Successful exciton confinement is achieved by implementing additional HTL and ETL, which have higher optical energy gaps than the exciton binding energy, between the emission layer and the transport layers. The organic layer that blocks electron transport into the HTL is called electron blocking layer (EBL) and the one which blocks hole transport into the ETL is called hole blocking layer (HBL). Well-known HBLs are oxidazole and triazol containing molecules, 1,3,5-tris(2-N-phenylbenzimidazolyl)benzene (TPBi) and 3-(4-Biphenylyl)-4-penyl-5-(4-*tert*-butylphenyl)-1,2,4-triazole (TAZ), and EBLs are fac-tris(1-phynylpyrazolato,*N*,*C2'*)iridium(III) (Irppz) and Iridium(III)bis(1-phenylpyrazolato,*N*,*C2'*)(2,2,6,6-tetramethyl-3,5-heptanedionato-*O*,*O*) ($ppz_2Ir(dpm)$) [39, 47, 48].

Furthermore, if doped charge transport layers with high conductivity are employed, the efficiency of OLED can be significantly enhanced. Furthermore, the driving voltage can also be reduced significantly to less than 3 V at 1,000 cd/m^2. The related mechanisms in the doped transport layer and the resulting effects in OLED were described by Walzer *et al.* [40], and they will be discussed briefly in section 5.2.1. The energy level diagram of a p-i-n OLED with exciton blocking layers is illustrated in Figure 2.6.

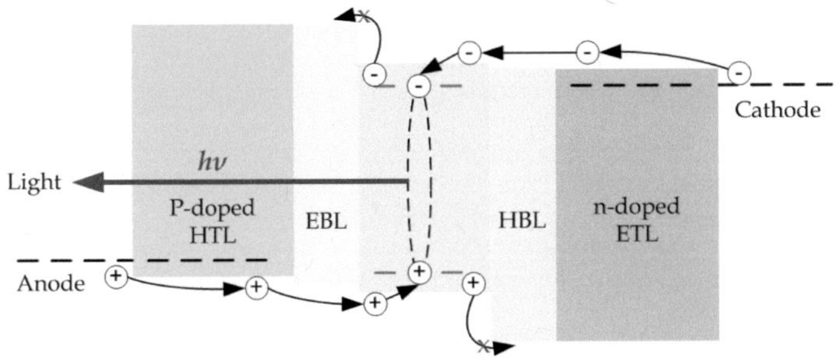

Figure 2.6: Schematic energy level diagram of p-i-n OLED illustrating the role of EBL and HBL. The phosphorescent dopant HOMO and LUMO levels are also indicated as dashed lines in the emission layer.

By changing electrode materials and substrates OLED can have various structures. So far, three structures have been reported as the emission directions: BEOLED, top-emission OLED (TEOLED), and transparent OLED (TOLED) [49-52].

BEOLEDs have been widely used in displays and lighting applications. Initial displays were commercialized with passive-matrix OLED (PMOLED) displays because the manufacturing processes are simple. In PMOLED, the individual pixels are operated by the electric field induced by the overlap of ITO columns (anodes) and metal rows (cathodes). Therefore, it possesses technical limitations to high resolution and large area display (typically up to three inches maximum) due to significant voltage drop through the electrodes and external integrated circuits (ICs), etc. However, PMOLEDs can still be found in MP3 players and sub-displays for cellphones. In order to solve those problems, thin film transistors (TFTs) are addressed to operate each pixel like TFT-LCD. OLED with this operation method is called active-matrix OLEDs (AMOLEDs). AMOLEDs are more expensive, and much more difficult to manufacture, but can be used for larger displays with no crosstalk, continuous excitation, low power consumption, continuous cathode configuration, and on-substrate IC.

Unlike the LCD where each pixel can be voltage-switched using a single addressing transistor, the OLED is current-controlled. The active devices need thus high carrier mobility in order to deliver charge carriers into the OLED, and they are fabricated using poly silicon (p-Si) on glass. The AMOLED circuit design faces with two primary issues: rather low electron mobility and large variations in electrical properties in poly-Si, which are associated with p-Si crystal growth and grain size. These two problems result in the variation of transistor threshold voltage leading to an uneven current delivery to each pixel and, consequently, luminance non-uniformity. Several techniques were studied for a crystallization of p-Si, such as low temperature solid phase crystallization, excimer laser annealing, and metal seeding [53], which have still problems like low throughput, high cost, and the presence of residual metals. Instead, the diode current in the emitting state can be continuously retained by means of two or more transistors which act as a drive transistor and address transistor, respectively [54].

In a two transistor circuit, the drive transistor drives the current for the OLED, and is operated in saturation to overcome the OLED threshold variation. The address transistor acts as a switch to sample and hold a voltage on the gate of the drive transistor. The major cause of luminance non-uniformity is the variation in the drive transistor. This transistor has a high level of output characteristic variation due to the nature of the polysilicon grain growth. A four-transistor circuit has been therefore proposed to reduce the effects of the transistor threshold voltage variation, resulting in a four-fold improvement in luminance uniformity as compared to the basic two-transistor circuit [55].

The power consumption of OLED displays is proportional to the number of driving pixels. In the AMOLED display, particularly, the internal luminance is governed by the aperture ratio at each pixel. However, the multi-transistor pixel circuits used for controlling each pixel in a BEOLED structure occupy quite a large space and result in a low aperture ratio (~40 %) and, consequently, shorter lifetime. In order to improve the aperture ratio, an inverted electrode structure was proposed, so-called TEOLED, which has a reflective anode as a bottom electrode and a semitransparent cathode as a top electrode. The schematic diagrams of BEOLED and TEOLED are shown in Figure 2.7 illustrating the luminance difference.

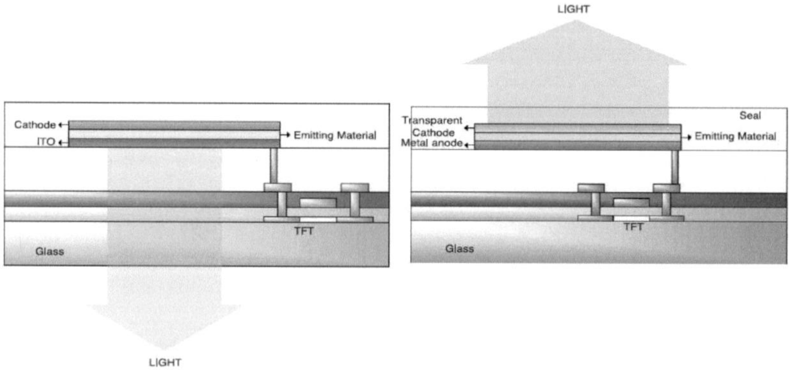

Figure 2.7: Structure of the BEOLED (left) and TEOLED (right). Reprinted from Ref. [56].

As semitransparent top electrode, a thin metal layer (~15 nm) having a transparency of 40 ~ 50 % in visible range is used [57]. Bottom electrode is usually a highly reflective metal such as Ag and Cr. The charge injection from the bottom electrode depends on the energy barrier that is determined by the workfunction of the electrode. By the two metal electrodes at both sides of the organic layers, an optical microcavity structure is formed which can enhance the performance of OLED [58]. The microcavity is optimized by varying the thickness of the organic layers, especially the transport layers, in order to produce a standing wave mode inside the cavity. The spontaneous emission intensity is then enhanced in the direction that is normal to the cavity axis. TEOLED is also applicable for opaque substrates like metal foil, which is more reasonable than glass substrate in the fabrication costs. Recently, much progress has been achieved to fabricate OLEDs on such flexible metal foils [59, 60].

If both electrodes are transparent, transparent lighting devices can be realized which enable OLED to address information device in automobile or aircraft windows. Thin metals are not favored due to their low transparency, but TCOs can sufficiently fulfill the requirements for high transparency. Since TCO depositions are usually performed by a magnetron sputtering method, the deposition of TCOs on organic thin films has been a great challenge. However, highly transparent (~ 80%) and efficient (> 50 cd/A) transparent OLED (TOLED) have been reported [61].

2.2.1.2 Metals for reflective electrode

The top metal electrode for bottom emission OLED is required to exhibit high reflectance in visible range for a better optical outcoupling and to have a good conductivity. Two materials, Al and Ag, are favored for these purposes.

Al is a soft, durable and light metal. Al is a nonmagnetic material and soluble in water in certain forms. However, its corrosion resistance is excellent due to a thin surface layer of Al oxide that is formed when exposed to air or humidity. Also, Al is one of the few metals that have silvery reflectance in a fine-powdered or thin film form.

As shown in Figure 2.8, Al has high reflectance in the 200–400 nm (ultra-violet) regions, while in the 400–700 nm (visible) and the 3,000–10,000 nm (far infrared) range the reflectance is slightly lower than Ag. Al has also a good bulk electrical conductivity of 2.7 μΩ.cm. Its melting point is 660.32°C, so it is more easily vaporized than Ag in high vacuum (< 1×10^{-6} mbar). Ag has a higher bulk electrical conductivity of 1.6 μΩ.cm and extremely good reflectance, but it is a noble metal. These characteristics enable Al to be the most frequently used electrode material in OLED fabrication.

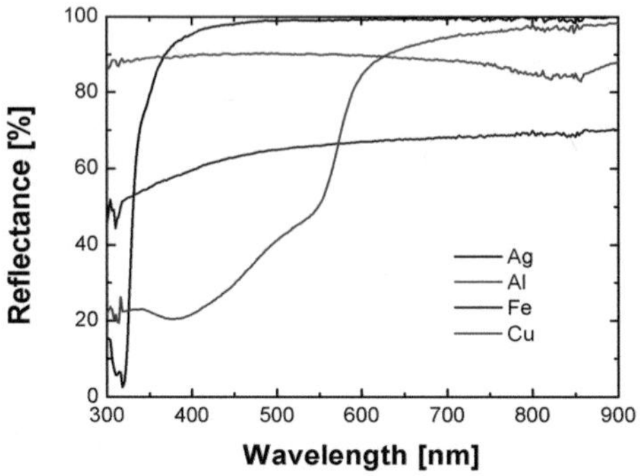

Figure 2.8: Spectral reflectance curves of 100 nm thick Ag and Al films deposited by PK500 sputter cathode in the Inline machine [62, 63]. Iron (Fe) and copper (Cu) reflectance curves were obtained from electropolished metal plates.

When Al or Ag is deposited onto an intrinsic organic layer, a Schottky contact is formed with a certain barrier height. The barrier height results in a lower current injection into OLED, so alkali metal-oxide compounds and lithium fluoride (LiF) were used between the organic layer and Al as a bilayer [64, 65]. The bilayers exhibited better electron injection into the organic layer and, consequently, lower driving voltage and higher efficiency by a better balance of charge carriers in the emission zone. The mechanism for enhanced electron injection is explained by a dissociation of alkali metal oxides at the interface and resulting alkali metal reaction with Al. For a better thermodynamic stability at the interface, LiF is preferred with a comparable performance.

2.2.1.3 Transparent conducting oxide for transparent electrode

There are many kinds of TCOs such as indium tin oxide (ITO), indium zinc oxide (IZO), and aluminum zinc oxide (AZO), etc. These TCOs are essentially used as a conducting electrode in every optoelectronic device such as TFT-LCD, LED, OLED and photovoltaic devices due to their high transparency in visible range and good electrical conductivity [66-68]. Besides, such electronic devices, TCOs can also be used for antireflection as well as antistatic coatings for color monitor or energy-efficient window [69, 70].

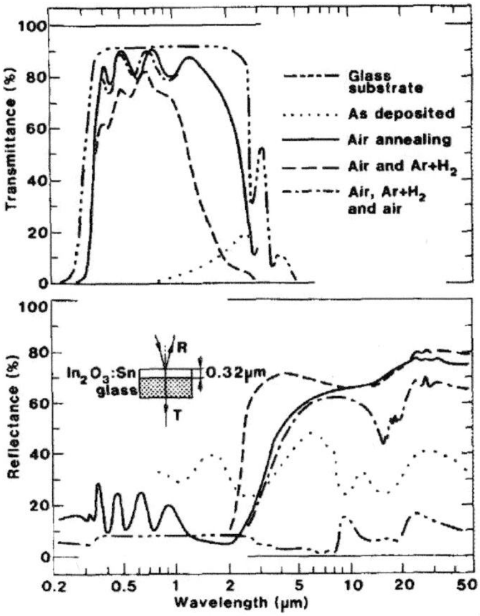

Figure 2.9: Optical characteristics of an evaporated In_2O_3:Sn film. Spectral transmittance (upper part) and reflectance (lower part) for a glass substrate and after coating on the glass with various post-treatment conditions. The graph was reprinted from Ref. [70].

The simultaneous existence of high optical transparency (more than 80%) in the visible region and low electrical resistivity (about 10^3 Ω·cm or lower) is usually not possible in an intrinsic stoichiometric material. Low transparency and good conductivity can be obtained in thin films at a thickness under ~30 nm. The only way to obtain good transparent conductors is to create electron degeneracy in a wide band gap (greater than 3 eV) oxide by incorporating non-stoichiometry and appropriate dopants. The mentioned TCOs are all such kinds of compounds. In the case of ITO, which has been the most frequently used transparent electrode, five to ten weight percent of SnO_2 is doped into In_2O_3 host lattice. The difference of doped or undoped material is usually understood by analyzing the optical properties. For that, the band structures of In_2O_3 and In_2O_3:Sn should be discussed.

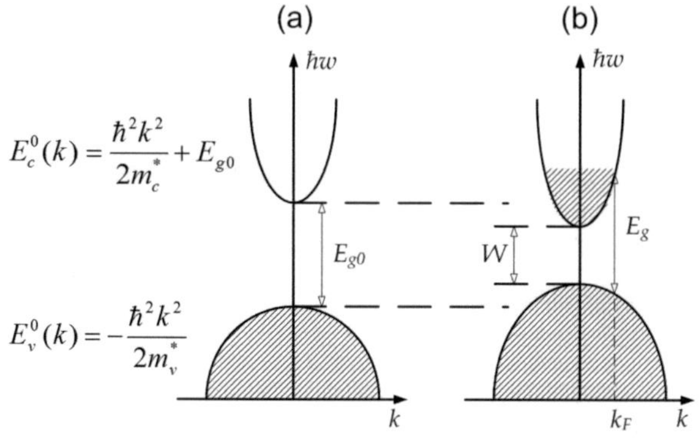

Figure 2.10: Figure (a) shows the assumed band structure of undoped In_2O_3 in the vicinity of the top of the valence band and the bottom of the conduction band. Figure (b) depicts the effects of Sn doping. A shift of the bands is apparent, and filled areas denote occupied states. Reprinted from Ref. [70].

Figure 2.10 shows the band structure of undoped In_2O_3. Parabolic bands are characterized by m_c^* for the conduction band and m_v^* for the valence band. Both bands are assumed to be isotropic. The direct semiconductor band gap of In_2O_3, denoted as E_{g0}, was reported to be 3.75 eV [71]. It was proposed that the conduction band is mainly from indium 5*s* electrons and the valence band is from oxygen 2*p* electrons. For the undoped material, the Fermi energy E_F is located in the middle between the energy bands. If a low density of donor atoms with a valence band slightly lower than the conduction band of a host material is introduced, donor states are formed below the conduction band of the host material. This is the general theory of the effect of doping, and E_F lies between the donor level and the conduction band minimum. With an increasing donor density, the donor states merge with the conduction band at a certain critical density n_c, and the magnitude is estimated from Mott's criterion [72, 73]

$$n_c^{1/3} a_0^* \approx 0.25 \tag{2. 2. 1}$$

The effective Bohr radius a_0^* is given by

$$a_0^* = h^2 \varepsilon_0 \varepsilon^M / \pi e^2 m_c^* \tag{2. 2. 2}$$

where ε^M is the static dielectric constant of the host lattice. With data for In_2O_3, $\varepsilon^M = \varepsilon^{In2O3}(0) = 8.9$ and $m_c^* \approx 0.35m$, a_0^* is approximately found as 1.3 nm [70]. Hence free-electron properties can be expected when $n_e > n_c \approx 6 \times 10^{18}$ cm^{-3}. Since normal TCOs show a carrier density above 10^{20} cm^{-3}, the free electrons contribute to the conduction. Above the Mott critical density, E_F is determined by the highest occupied states in the conduction band, and E_F is written as

$$E_F = (\hbar^2 / 2m_c^*) k_F^2 \tag{2. 2. 3}$$

where the magnitude of the wave vector at the Fermi energy is $k_F = (3\pi^2 n_e)^{1/3}$.

A partial filling of the conduction band as well as a shift in energy of the bands are shown in Figure 2.10 (b). The band gap will gradually shift towards the higher energy as the doping concentration, namely the electron density, is increased. Optical band gap is easily observed from the onset of absorption coefficient derived from the optical constants on energy. The absorption coefficient is given by

$$\alpha = 4\pi k / \lambda \qquad (2.2.4)$$

However, this method is approximately correct for direct allowed transitions to an empty parabolic conduction band, and is not valid for the heavily doped semiconductor. Since the lowest states in the conduction band are filled, transitions can only take place to energies above E_F, as illustrated in Figure 2.10 (b). However, quantum-mechanical transitions between states in a filled parabolic valence band and a partially filled parabolic conduction band are enabled by band-gap shift and perturbation.

In order to model the shift, the band structure should be discussed. As indicated in Figure 2.10 (a), the dispersions for the valence and conduction bands in the undoped In_2O_3 lattice are written as

$$E_v^0(k) = -\hbar^2 k^2 / 2m_v^* \qquad (2.2.5)$$

$$E_c^0(k) = E_{g0} + \hbar^2 k^2 / 2m_c^* \qquad (2.2.6)$$

where k is the wave vector and superscript 0 denotes unperturbed bands. By the introduction of dopants, the band gap E_g is widened as shown in Figure 2.10 (b). But electron-electron and electron-impurity scattering tend to decrease the gap, denoted as W which is the energy difference between the top of the valence band and the bottom of the conduction band in the doped material [74, 75]. When the blocking of the lowest states in the conduction band is considered first, neglecting scattering effects, then

$$E_g^0 = E_{g0} + \Delta E_b^{BM} \qquad (2.2.7)$$

where ΔE_g^{BM} is the band-gap shift given by Burstein-Moss (BM) theory [76, 77],

$$\Delta E_g^{BM} = (\hbar^2 / 2m_{vc}^*)(3\pi^2 n_e)^{2/3} \qquad (2.2.8)$$

with the reduced effective mass

$$(m_{vc}^*)^{-1} = (m_v^*)^{-1} + (m_c^*)^{-1} \qquad (2.2.9)$$

Eq. (2. 2. 8) yields a band-gap change proportional to $n_e^{2/3}$, which is in good agreement with experiment results [78-80].

For the variation of bandgap structure, there can also be other effects by the local deformations of the lattice resulting from randomly distributed Sn atoms or defects. This mechanism is responsible for the Urbach tails in the absorption spectrum that low-energy parts of band gaps look almost linear [81]. Another approach has been also chosen to treat the broadening of the band gap following Redfield, who described tailing through a Franz-Keldysh effect of microfields caused by the ionized defects such as Sn^{4+} in In_2O_3 [82, 83].

The heavy doping affects electrical properties of TCOs as well as optical properties. The high conductivity of pure oxide films is due to the high carrier (mostly electron) concentration because the mobility in TCO films is considerably lower than that in the bulk material. The high electron concentration is attributed to a deviation from stoichiometry (or defect structure) and a doping effect leading to controlled valence semiconductors [84, 85]. The stoichiometric deviation is due to anion vacancies or excess interstitial cations. In case of SnO_2, by combining x moles of SnO with 1-x moles of SnO_2 the non-stoichiometric oxide $Sn^{4+}{}_{1-x}Sn^{2+}{}_xO^{2-}{}_{2-x}$ is formed, which contains x moles of oxygen vacancies with x moles of Sn^{2+} free to donate 2x moles of electrons for conduction. Oxygen vacancy control is therefore the key to increase n-type conductivity or produce a hole which acts as a trap in the n-type semiconductor and decreases the conductivity. ITO is represented as $In_{2-x}Sn_xO_{3-2x}$, and substitutional tin and oxygen vacancies contribute to the high conductivity [86].

A quantitative measure of performance of a TCO can be written in terms of the ratio of the electrical conductivity to the optical absorption coefficient or,

$$\sigma_e / \alpha = 1[R_s \ln(T + R)]^{-1} \qquad (2.\,2.\,10)$$

where T and R represent the total transmittance and reflectance of the material, and R_s is the sheet resistance (=resistivity/film thickness) [87]. Electron conductivity σ_e is derived from Ohm's law. Electron drift velocity in a medium is proportional to the applied field. When the electrons are forced by an external electric field, free electrons feel a force of magnitude eE and start to drift. However, the electrons collide with lattice ions, and the collisions hinder an electron movement. If τ is the average time between collisions, the conductivity of electron is defined as

$$\sigma_e = n_e e^2 \tau / m_e^* \qquad (2.\,2.\,11)$$

where n_e is the electron density, e is the electron charge and m_e^* is the effective electron mass [88]. This figure-of-merit is used to assess the relative performance of different TCO materials and is wavelength dependent. It should be noted here that the factors in Eq. (2. 2. 10) are highly dependent on process conditions, equipment type and the resulting microstructure and defect chemistry of the material. Thus, a highly optically scattering material, due to the presence of a polycrystalline grain structure, will have both decreasing transmission and higher resistivity due to a decrease in charge mobility arising from scattering of charge carriers at grain boundaries.

The mechanism responsible for the film mobility in TCOs is mainly scattering of carrier charges in the film. Main scattering mechanisms in the TCO films are ionized and neutral impurity scattering, thermal lattice vibration scattering and grain boundary scattering [89]. Since the concentration and the scattering cross section of the neutral impurities are smaller than that of ionized impurities, the neutral impurity scattering is neglected. The reciprocal mobility of the films is expressed as

$$\frac{1}{\mu} = \frac{1}{\mu_i} + \frac{1}{\mu_l} + \frac{1}{\mu_g} \qquad (2.\,2.\,12)$$

where μ_i, μ_l, and μ_g are mobility of ionized impurity, lattice vibration, and grain boundary scattering, respectively.

Grain boundaries in polycrystalline films are formally assumed to contain lattice defects-induced trapping states [89]. These states are thought to compensate some fractions of charge carriers of ionized dopants, creating potential barriers across the depletion region. These barriers can hinder carrier motion from one crystallite to another. The width of the depletion layers and the grain size of the crystallites are correlated with each other to explain the carrier motion, and this model has been widely used for the TCO films. The grain boundaries between crystallites can contain oxygen which can capture electrons from the conduction band, making the grain boundaries negatively charged [91]. As a result, potential barriers are generated, and they result in a strong scattering of conduction electrons. R. L. Petritz described a model of the grain boundary scattering-dependent mobility based on Maxwell-Boltzmann statistics as

$$\mu_g = AT^{-1/2} \exp[-(eV_B / kT)] \qquad (2.\ 2.\ 13)$$

where V_B is the potential barrier between the grain boundaries, $A = el/(2\pi m^* k)^{1/2}$ is a constant for a sample [90]. For TCO films such as undoped SnO_2 or antimony-doped SnO_2, V_B is usually found at around 0.02~0.03 eV [92, 93]. The thickness of depletion layers at grain boundaries can be calculated using the equation

$$d = \sqrt{\frac{2\varepsilon_0 \varepsilon_r V_B}{qn}} \qquad (2.\ 2.\ 14)$$

If the carrier density n=6×10^{20} cm^{-3} is substituted together with other known parameters to Eq. (2. 2. 14), the calculated thickness of the potential barriers is only a few Å. In case of the potential barrier with such small thickness, a tunneling current between grains may contribute to the transport of charges across grain boundaries dominantly.

2.2.2 Electrical characteristics of OLED

The current (I)-voltage (V) characteristics of OLED are governed by three current regimes, which are based on the charge transport mechanisms described in section 2.1.3. In Figure 2.11, *I-V* characteristic of a p-i-n OLED is shown, and a forward *I-V* characteristic is included in the inset.

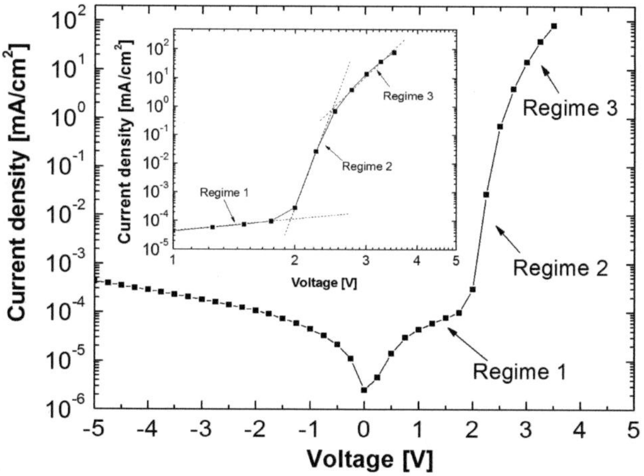

Figure 2.11: *I-V* characteristic of a typical p-i-n OLED. Inset is the forward *I-V* characteristic on voltage in log scale. Three current regimes with different orders of voltage are observed.

Regime 1: The current at low voltage is proportional to the voltage, $I \propto V$. The current follows Ohm's law and is dominated by thermally generated charges [94], thus

$$J = n_0 q \mu \frac{V}{d} \tag{2. 2. 15}$$

where n_0 is the equilibrium electron density, q the electronic charge, μ the charge mobility and d the thickness of the organic layer. The ohmic current density in Figure 2.11 is much higher than that of OLEDs which have only intrinsic organic layers [35]. It attributes to more free charges in the doped transport layers. This current also represents leakage originating from particle contamination, sharp surface structures of substrate, shortcuts, etc. Leakage currents show different behaviors with respect to the driving voltage. They show largely three types: The first type is the typical Ohmic current that the current increases with the voltage with a power of one, which obeys Eq. (2. 2. 15), as depicted in Figure 2.12 (a). This leakage current is explained by a complete short circuit of two electrodes as a result of the sharp surface structures or a metallic particle contamination. The level of this current can be then influenced by the number and the size of the short circuits. The second type is an exponentially increasing current on the driving voltage to a square or a larger order, as depicted in Figure 2.12 (b). This current is regarded as an incomplete short circuit. As a result, a much higher electric field can be induced through the organic layer sandwiched between two electrodes, and the charge conduction may appear like a trap-free conduction. In some cases, the leakage current fluctuates severely like Figure 2.12 (c), which may result from mobile metallic atoms within the organic layers under an increasing electric field. These leakage current forms will be shown in section 6.

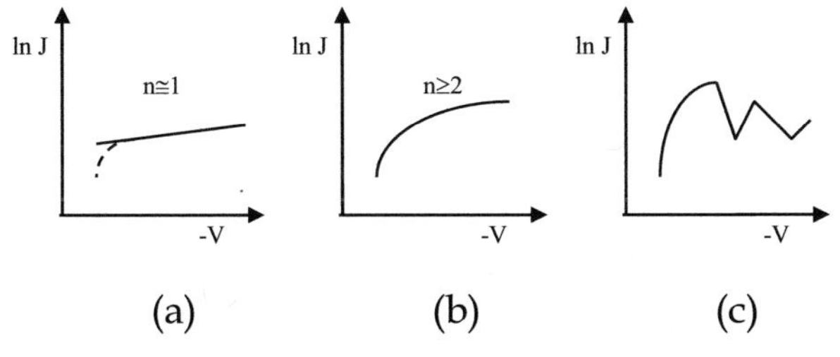

Figure 2.12: Three types of leakage currents in OLED after sputtering; (a) Ohmic, (b) metal-induced trap-free and (c) fluctuating leakage currents.

Regime 2: As the voltage is increased further, most of the traps are filled and the current increases on the voltage with an order of more than approximately eight. This current is TCLC, which was derived in Eq. (2. 1. 8), and is influenced by the trap density. Charge carriers are recombined in the emission zone, and the luminance is influenced by the variation of the current within the current regime.

Regime 3: At the higher voltage, the current reaches trap-free region, and the transport mechanism is space-charge limited. The current shows a square-law dependence on the voltage and is thus SCLC derived in Eq. (2. 1. 9).

2.2.3 Definition of OLED efficiencies and determining factors

Three kinds of efficiency are used to evaluate OLEDs: External quantum efficiency, power efficiency, and current efficiency. The external quantum efficiency η_{ext} is one of the most critical figures of merit in OLED lightings and displays. There are several processes which influence the external quantum efficiency.

After the charge carriers are injected from both electrodes, the ratio between the electron and hole currents determines the recombination efficiency γ, in other words, the charge carrier balance factor. It can reach almost one, if both currents injected are equally balanced and the emission zone is surrounded by the exciton blocking layers.

When the electrons and holes form excitons, the ratio of singlet and triplet excitons is determined by spin statistics. As discussed in section 2.1.2, singlet and triplet excitons are generated with a ratio of 25 % and 75 %, respectively. Therefore, the singlet excitons recombine in fluorescent emitters radiatively with a total radiative recombination efficiency η_r of 25 %. As seen in Figure 2.4, the intersystem crossing of the singlet excitons to the triplet excitons in a host-phosphorescent emitter system is non-radiative, but a very efficient process. As a result, the singlet excitons are also harvested for the phosphorescent emitters, and the radiative recombination efficiency η_r can reach 100 %. Accordingly, the fraction of excitons η_{ST} that decay radiatively is given as 0.25 and 1 maximum for fluorescent and phosphorescent excitons, respectively. The internal quantum efficiency is determined by the three parameters, following:

$$\eta_{\text{int}} = \gamma \times \eta_r \times \eta_{ST} \tag{2. 2. 16}$$

The external efficiency η_{ext} is obtained from η_{int} by including the outcoupling efficiency η_{out}. In a typical BEOLED, a maximum of 20 % of the generated light can be coupled out [95]. This is due to the different refractive indices of glass substrate and air, and η_{out} is given by

$$\eta_{out} = \frac{1}{2}\frac{n_{air}^2}{n_{glass}^2} \approx 0.22 \tag{2. 2. 17}$$

which is simply estimated with neglecting interference effects, different refractive indices of bottom TCO electrode and organic layers, and plasmon loss channels, etc. In total, η_{ext} is expressed as

$$\eta_{ext} = \gamma \times \eta_r \times \eta_{ST} \times \eta_{out} \tag{2. 2. 18}$$

The power efficiency (PE) η_p and the current efficiency (CE) η_c are also the performance index of OLEDs in photometric (lm/W) and radiometric (cd/A) unit, respectively. In general, η_p is used for a lighting application because it defines the device power consumption at a specified brightness, while η_c is frequently used for display devices.

η_p is defined as

$$\eta_p = \frac{L_p}{IV} \tag{2. 2. 19}$$

where L_p is the total light output in lumens, I current, and V voltage.

η_c is defined as

$$\eta_c = \frac{I}{JA} \tag{2. 2. 20}$$

where I is the luminous intensity, J current density, and A emitting area.

3 Process techniques for OLED fabrication

This section provides brief introductions of various process technologies for OLED fabrication. Depending on the materials to deposit, the process and method can be varied. In this study, the deposition of the organic and electrode materials was mostly performed by vacuum evaporation, the so-called physical vapor deposition (PVD), so vacuum processing techniques are discussed in detail. The motivation of this study arose from the resulting phenomena in the metal evaporation, which are undesirable for OLED. They will be described in each corresponding section.

3.1 Organic material deposition

3.1.1 Thermal evaporation

Most small molecule organic materials are deposited on a given substrate by means of thermal evaporation in vacuum [2, 40]. Vacuum evaporation is a basic method for thin film deposition. A source material is evaporated by the Joule heating to its vaporizing temperature at a given vaporizing pressure. The ejected atoms or molecules are then condensed on the substrate.

Atoms or molecules evaporating from a cell have kinetic energies according to their temperatures. It is important to know the directional distribution of the evaporating molecules to obtain a good homogeneity of film thickness. In general, the evaporation occurs according to Knudsen's cosine law [96], and the concept is successfully applied to point sources, which are used for the evaporation of organic materials.

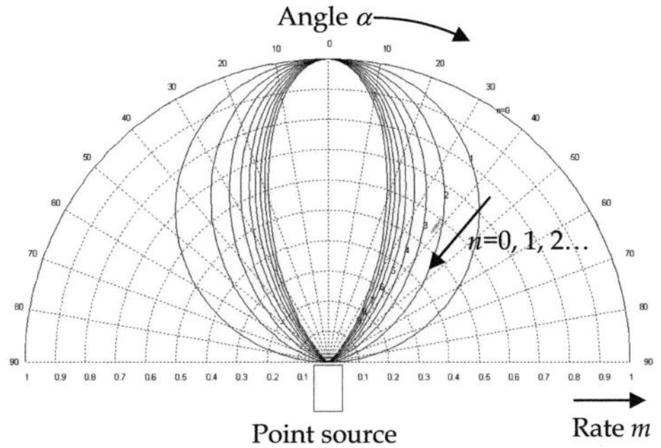

Figure 3.1: Rate distribution of evaporating material from a point source.

The evaporating rate m_α at an angle α is given as

$$m_\alpha = m_0 \cos^n \alpha \qquad (3.1.1)$$

where m_0 is the rate at the angle of 0° and the cosine exponent is the coefficient depending on the own property of the point source. When n=1, the evaporation source is called a Knudsen-cell source. Most evaporators have vapor characteristics corresponding to n values between n=3 and n=5, and the resulting thickness profile is strongly influenced by the n value.

Depending on the source materials, several methods are applied to heat the source material, e.g. radiative heating by an electric filament, direct heating on a heating element, or local heating by a high energetic electron beam (15 keV). The vaporizing temperature of small-molecule organic materials ranges normally within 100~400°C. The radiative heating method is widely used for the deposition of organic thin films because the direct heating or the electron beam can result in the deterioration of organic materials.

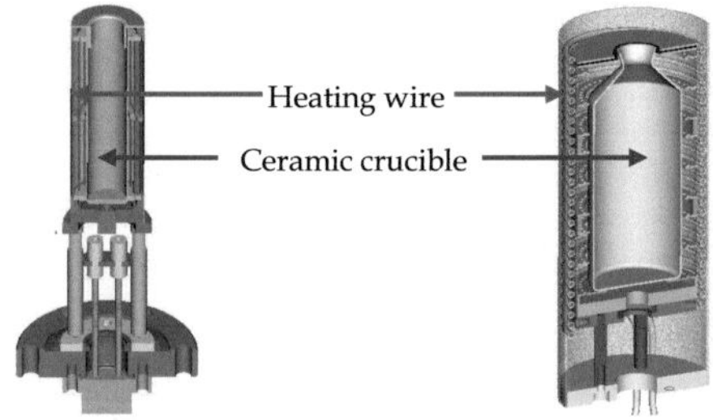

Figure 3.2: General structures of effusion evaporation sources for organic materials, with 10 cc crucible volume (left) and 135 cc large crucible volume (right). (with permission from Sunic system Co. Ltd.)

General structures of effusion evaporation sources for organic materials are illustrated in Figure 3.2. Organic materials are filled in a crucible made of boron nitride (BN), poly-boron nitride (p-BN) or ceramic (Al_2O_3), and a high-resistive metal wire or plate surround thereby. By applying a suitable power, the radiative heating is generated from the heating element and transferred to the organic materials via the wall of the crucible. When the temperature of the organic materials exceeds their vaporizing temperature, the molecules are effused through the hole of the crucible and condensed on a substrate.

The vacuum evaporation has a lot of merits, but the poor controllability and the reproducibility of the deposition rate is a critical problem in the evaporation equipments. Therefore, the vacuum evaporation essentially needs a thickness monitor consisting of a quartz crystal sensor and a controller for a real-time feedback of film thickness being deposited. The size and type of the organic evaporator are determined by a configuration of the process chamber or a substrate size to be covered. Detailed information will be described in the following equipment sections.

3.1.2 Organic vapour phase deposition (OVPD)

Organic vapor phase deposition is a specially designed deposition method for organic materials, and its technique stems from a metal-organic chemical vapor deposition (MOCVD). OVPD was invented by Burrows *et al.* at Princeton University in 1995 as a novel method for the deposition of organic thin films [97]. OVPD uses an inert gas as a carrier of vaporized organic materials. Its advantage is the easy control of the deposition rate and the morphology of the film by changing the gas flow rate and the source temperature. Compared to the thermal evaporation, a more uniform film thickness can be achieved by the help of the carrier gas. So a closer source-to-substrate distance than that of the thermal evaporation is possible so that a material utilization can be improved. OVPD is being used in various applications of organic electronics, e.g. OLED or organic photovoltaic (OPV). A representative manufacturer is Aixtron AG, Germany. If the OVPD is combined with a cluster type equipment, the material utilization for large area OLEDs can be significantly improved [98].

3.1.3 Solution processing

Solution processing is to coat soluble materials on a given substrate in ambient. Among them, spin-coating is a very popular method to coat a solution onto a flat substrate. This technique is widely used in semiconductor industry for a photo resist (PR) coating on a substrate like a Si wafer. The substrate rotates at a high speed in order to uniformly distribute the coated solution. The thickness of the film can be controlled by the rotation speed, and it also depends on the viscosity and density of the solution. In OLED applications, a polymer material in solution can be coated with this technique in atmosphere. Representative polymer materials are poly(3,4-ethylenedioxythiophene) (PEDOT:PSS) and polyanilin as a bottom electrode [99, 100], derivatives of poly(*p*-phenylene vinylene) and polyfluorene as an emission layer [101]. Those materials are coated by either spin-coating or inkjet printing for fine patterns. However, the lifetime of the light emitting-polymers is still a bottleneck, and a major reason is low material purity. On the other hand, a successful result in the lifetime of the OLED on polymer anode was also reported by Fehse *et al.* that showed a comparable lifetime to that of the OLED on an ITO anode after annealing [99]. Polymer light emitting diodes (PLEDs) and other polymer-based organic devices are very cost-effective in production, but a breakthrough is still necessary in material development to improve their lifetime and performance. Spin-coating will then become a favorable method for a small device size, and a slit coater should be implemented due to a large centrifugal force of the spin-coater for a large substrate during high-speed rotation. The slit coater has been already used in TFT-LCD for PR coating. The slit coater is suitable for a single rectangular-shaped layer deposited on a large area device such that it can be useful for the mass-production of a large area OLED lighting or OPV applications.

3.2 Top electrode deposition

In this section, two representative evaporation methods used frequently for metal deposition are introduced. Evaporation is a comparatively easy process for coating purpose, but it has several problems in the stability of the heating element, deposition rate controllability, strong radiation, and scaling to large area substrate. These arguments will be discussed in the following subsections. The motivation of this study in fact arose from those listed problems that have been regarded trivial on the lab level but can be critical in the large area OLED mass-production. The problems can result in irregular maintenance term for the metal deposition so that equipment uptime is significantly influenced. Thus, magnetron sputtering was applied for this study in order to develop an alternative deposition method because the sputtering system can improve those problems. The principle of magnetron sputtering will be discussed in detail in section 4.

3.2.1 Thermal evaporation

For OLED fabrication, the thermal evaporation technique is generally used for a top electrode deposition on organic layers in spite of a thermal radiation from a heating element. It is mainly due to the low kinetic energy of thermally vaporized species in high vacuum. However, a restriction in source material selection exists because the vaporizing temperature must be low enough not to increase the substrate temperature over the glass transition temperature of underlying organic layers by the thermal radiation from the heating element. Not only the thermal load effect, but also the work function of the metal is a very important factor because the energy barrier between the deposited metal and the HOMO/LUMO levels of the organic layer, which make contact, determine the injection property of charge carriers. Thus, a rare earth metal with low work function like Yb can be used [102], or alkaline earth metals, e.g. Ba, Ca or Mg:Ag, are used for a contact with an electron transport layer [103, 104].

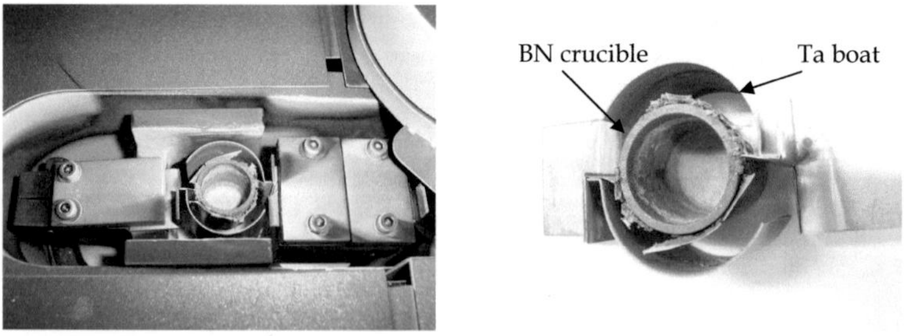

Figure 3.3: Metal evaporator installed in cluster equipment (Sunicel 200, Sunic system Co. Ltd.)

The evaporation temperatures of metals are in general higher than those of organic materials. Al is the most frequently used metal in OLED fabrication. Al is evaporated at over 677 °C and at a vapor pressure below 10^{-6} mbar which is a two or threefold higher temperature than that of organic materials. Thus BN crucibles surrounded by Ta or W boats are generally used to heat the Al source, as shown in Figure 3.3, because BN is more stable than ceramics at high temperature. However, such metal boats are supplied with high power in order to exceed the vaporizing temperatures of the electrode metals causing a gradual degradation of the boat materials. In particular, molten Al can easily wet the boat surface, resulting in serious destruction due to an extremely high local current.

It has also been reported that the liquid phase Al has good wettability to carbon-graphite materials [105]. In the Inline machine used (see section 0), the heating element for metal evaporation is made of a graphite plate so that Al evaporation is impossible because the Al coated on the heating element gives rise to a failure such as local corrosion.

3.2.2 Electron beam evaporation

Electron beam evaporation is also a kind of PVD which evaporates a source material by a bombardment of electron beam to cause a local heating on the source.

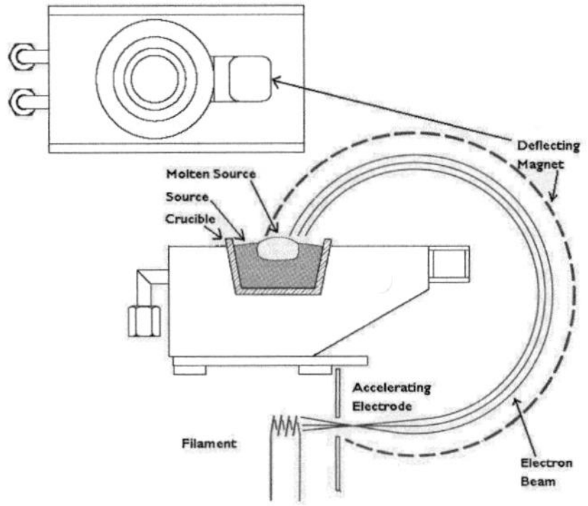

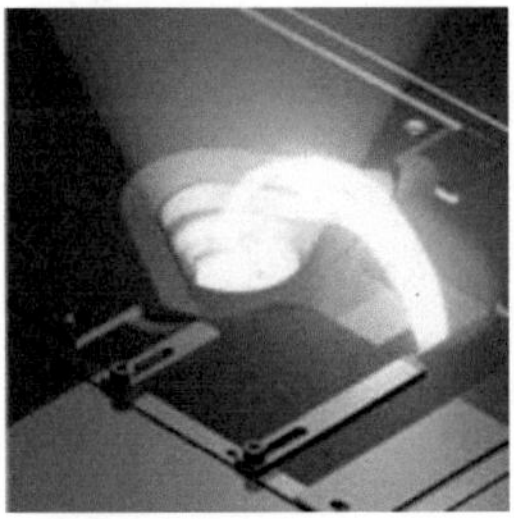

Figure 3.4: A schematic illustration of the electron beam evaporator (left) and a photograph of a real electron beam evaporator in operation (right). Photograph was taken from 'Oxford vacuum science Ltd., UK'.

A schematic diagram is shown in Figure 3.4. A tungsten filament inside the electron beam gun is heated to emit thermal electrons. These electrons are deflected and accelerated towards the source material by means of magnetic and electric fields. When the electron beam strikes the source surface, the kinetic energy is converted to thermal energy by the impact. This energy vaporizes the source material. In order to accelerate the electrons, the source material is supplied with a high DC voltage ranging from 10 to 30 kV. The deflection is accomplished by permanent magnets installed inside the electron beam gun. The magnets should be cooled for a protection from the heat generated.

By using the electron beam, the thermal energy delivered to the substrate can be considerably reduced because it induces a quite localized heating and results in a high deposition rate. However, in case of metal evaporation, X-ray and secondary electrons are emitted by the highly energetic electron impact on the metal [106, 107]. It was reported that they generate significant trap charges within semiconductor and interfaces of a multi-layer structure. Thus, the electron beam evaporation is usually avoided in OLED fabrication.

3.3 Vacuum equipment for OLED fabrication

Since small molecule organic materials were used in this study, high-vacuum equipment is only discussed. There may be novel equipment types, but two representative types are described which are widely used for OLED fabrication in laboratories and industries.

3.3.1 Cluster type equipment

Cluster equipment consists of several process chambers around a transfer chamber that has a robot handling system to transport substrates to each process chamber. The substrates are processed subsequently in each process chamber and continuously transported to other clusters. This type was designed for a high throughput in semiconductor mass production and is used nowadays in OLED as well as TFT-LCD production. Several manufacturers (Sunic system (Korea), Tokki (Japan), ANS (Korea), and Ulvac (Japan)) are providing the cluster type equipment from laboratory to mass production level. An example of fully automatic cluster equipment with three clusters is illustrated in Figure 3.5.

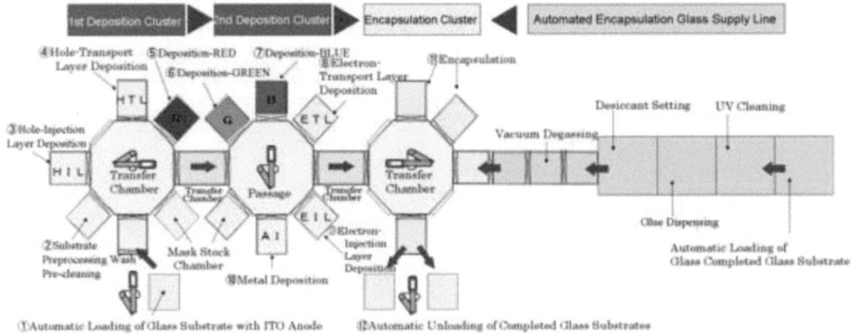

Figure 3.5: Schematic illustration of full-automatic cluster equipment for mass production. Image was taken from the internet homepage of Tokki Co., Japan.

3.3.2 Inline type equipment

An equipment type with the process chambers installed in a line is called Inline equipment. In the Inline equipment, substrates are loaded and processed continuously without stop in every chamber, so a very high throughput is accomplished. Furthermore, there is in principle no limit in equipment size if substrate production has no limit. As a film deposition source, a linear magnetron sputter system has been used with various cathode and magnetron forms, such as a static and dynamic magnetron or a rotating target.

Inline type vacuum equipment is used mostly for high speed and functional deposition on large substrate with high throughput. Those advantages are enabled by the nonstop process: the substrate loaded into the first vacuum chamber is transported to the next process chambers and deposited with thin films. In the meanwhile, the next substrate is loaded into the first vacuum chamber and follows the earlier substrate, which enables a continuous substrate flow. This procedure is repeated until all the substrates are processed and leave the unloading chamber.

Inline type equipment can also be designed with horizontal and vertical chamber configurations. The implementation of the vertical design yields two preferred advantages to the horizontal design. The one is less particle contamination, and the other is that large area glass can avoid bending downwards.

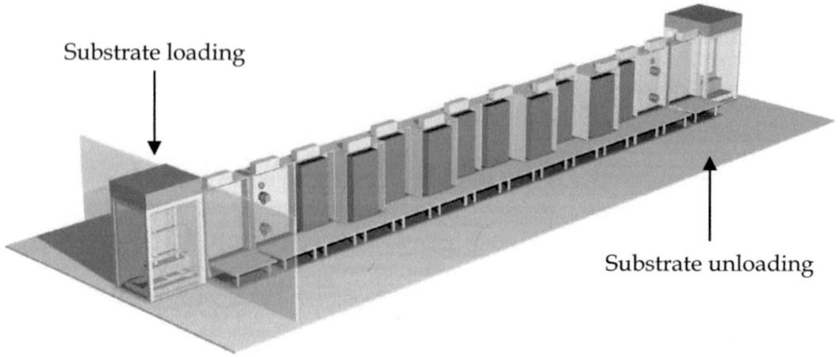

Figure 3.6: Schematic illustration of Inline type equipment. It consists of several process chambers, substrate load and unload chambers. Reprinted from Ref. [108].

There are many manufacturers for inorganic material depositions, but the number of manufacturer is quite small for organic material deposition. A representative for OLED is Applied Materials GmbH & Co KG, Germany.

4 Principle of magnetron sputtering and its applications

This section provides an introduction of magnetron sputtering and its applications, especially on thin film deposition and applications in the OLED field. Because the object of this study is sputtering deposition on organic layers, it is necessary to discuss the working principle of magnetron sputtering. In the first subsection, the fundamental of plasma is discussed which is essential for the understanding of sputtering process. In particular, only a weakly ionized plasma discharge where the working range of magnetron sputtering belongs to is described on the basis of the book from M. A. Lieberman [109]. In the second and third subsections, the mechanism of physical sputtering is introduced, and the various power sources for magnetron sputtering are briefly discussed. In the fourth subsection, a special sputter method, a FTS system, is introduced which is reported to be a low-damage sputter method. There have already been several attempts to deposit top electrode on organic layers, but critical degradation in OLED were unfortunately reported. The reported damage sources are summarized in the last subsection in order to realize the realization of damage-free sputter deposition on organic layers.

4.1 Plasma, a weakly ionized glow discharge

Plasma has a wide application spectrum, e.g. lighting, surface treatment, semiconductor processes, etc. Plasma is usually defined as a fourth state of matter, and it is considerable to describe the physics of plasma discharge for this study.

A solid substance in thermal equilibrium passes into a gaseous state at a sufficiently high temperature through a liquid state. If the temperature is increased, the gaseous atoms or molecules are decomposed into freely moving charged particles (ionization), and this state is called plasma. The charged particles are electrons and positive ions. In equilibrium, the plasma is characterized by a common charged particle density $n_e \approx n_i \approx n$ particles/m^3 at a temperature $T_e = T_i = T$. The energy required to form plasma from the solid state in thermal equilibrium ranges from approximately 4,000 K for easily ionized elements to 20,000 K for hardly ionized elements. In a vacuum, where gases exist, an energy supplied heats the gases, and the gases enter a plasma state through the ionization processes.

The plasma is a kind of collection of free charged particles moving in random directions. Therefore, they are electrically driven, and collision processes between charged particles and neutral gas molecules are important. There are many different kinds of plasmas, depending on the density of charge particles and the temperature of electrons. In general, low-pressure discharges are characterized by $T_e \approx 1 \sim 10$ V, $T_e \gg T_i$, and $n \approx 10^8 \sim 10^{13}$ cm^{-3}. These discharges are used for breaking source gases into positive ions and chemically reactive etchants, deposition precursors, etc. These particles also flow onto the substrate surface to react physically or chemically. The energy is then delivered to the substrate, for example in the form of ion bombardments. The electron temperature T_e is normally two orders larger than the ion temperature T_i, but the energy of ions bombarding the substrate can exceed substantially T_e because ion masses are incomparably larger than the electron masses. This is also the reason why ion bombardments are required for the chemistry at the substrate because the bombardments can increase chemical reaction rates and sputter materials physically from or into the surface.

The ions are accelerated across a thin sheath region where the plasma and substrate meet. Since this sheath region plays a main role in sustaining and processing discharges, the detailed sustaining mechanism will be discussed later. Electrons ionize or dissociate gas to create positive ions or free radicals, etchant atoms, and deposition precursors. Typically, T_e is less than the threshold energies for dissociation and ionization of the gases and molecules. Then, one may suspect how such dissociation and ionization occur. It is because electrons have a wide distribution of energies. The electron distribution function $g_e(E)$ is shown in Figure 4.1. E_{diss} and E_{iz} are the dissociation energy and the ionization energy respectively. Electrons having energies below E_{diss} and E_{iz} cannot dissociate or ionize the gas. The electron distribution appears like a Maxwellian function, but a tail distribution exits over the bulk electron temperature T_e. The tail distribution may be enhanced above a Maxwellian by electron heating and electron-neutral collision processes.

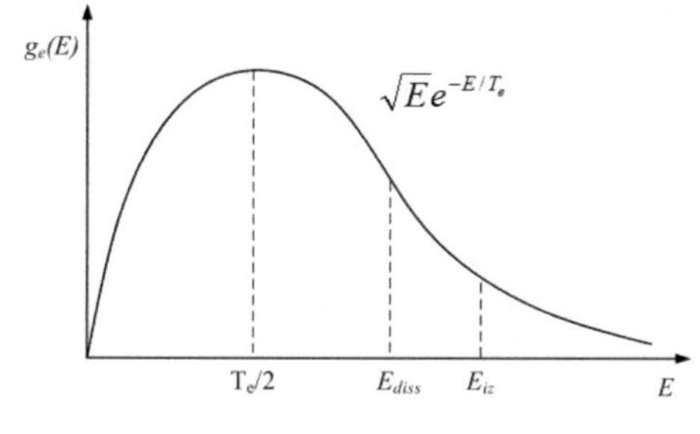

Figure 4.1: Electron distribution function in a weakly ionized discharge.

When the plasma meets the reactor wall surface, a thin positively charged layer is generated, the so-called sheath. It should be noted that the electron thermal velocity $(eT_e/m_e)^{1/2}$ is at least 100 times the ion thermal velocity $(eT_i/m_i)^{1/2}$ because $m_e/m_i \ll 1$ and $T_e \gg T_i$. The formation of the plasma sheath is schematized in Figure 4.2. A plasma of width l with $n_e = n_i$, confined initially between two grounded walls is taken into account. Potential at the wall is Φ=0. Since the net charge density $\rho = e(n_i - n_e)$ is zero, the electric potential Φ and the electric field E_x is zero everywhere (Figure 4.2 (a)). Hence, the fast-moving electrons are not confined and will be lost to the walls quickly. On a very short timescale some electrons near the walls are lost, leading to the situation shown in Figure 4.2 (b). Thin ($s \ll l$) positive ion sheaths are formed near to each wall, which exhibit $n_i \gg n_e$. The net positive ρ within the sheaths leads to a positive potential profile $\Phi(x)$ within the plasma, falling sharply to zero near both walls. This is why the plasma potential is always positive relative to the chamber walls, resulting in a confining potential valley for electrons and a hill for ions because the electric fields within the sheaths direct from the plasma to the wall. Therefore, the force $-e\vec{E}_x$ acting on electrons is directed into the plasma, and the electrons travel toward the walls back into the plasma. In contrast, ions from the plasma entering the sheaths are accelerated into the walls. If the plasma potential with respect to the walls is V_p, then one expects that $V_p \sim$ a few T_e in order to confine most of the electrons. The energy of ions bombarding the walls is then $E_i \sim$ a few T_e. Hence the division of discharges into bulk plasma and sheath regions is a very important factor that applies to all discharges. The bulk plasma dynamics are described by a diffusive ion loss at high pressures and by free-fall ion loss at low pressures. In the positive space charge sheaths, high fields exist, leading to dynamics that are described by various ion space charge sheath laws, including low-voltage sheaths and various high-voltage sheath models. Both plasma discharge dynamics are described with collisionless and collisional modifications. It is therefore necessary to discuss the collision processes of particles that exist in the plasma.

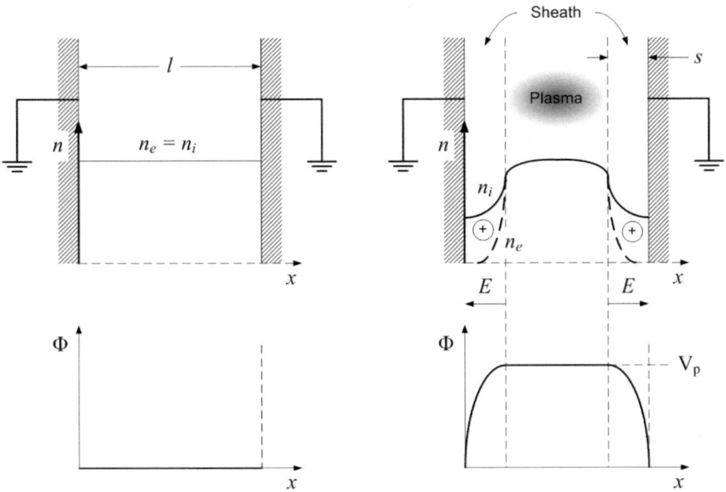

Figure 4.2: The formation of plasma sheaths : (a) initial ion and electron densities and potential; (b) densities, electric field, and potential after formation of the sheath

A collision is an event in which two or more moving particles interact with relatively strong forces on each other for a short time. By collision various phenomena may occur, for instance, single or multiple particles change their momentum or their energy, neutral particles can be ionized, and ionized particles can become neutral. Simple classical estimates of important processes in gas discharges are discussed, which are also essential processes in magnetron sputtering.

The fundamental quantity that characterizes the collision is a cross section of particles. To define this, a simple situation shown in Figure 4.3 is considered, in which a flux $\Gamma = nv$ of particles having the mass m, density n, and fixed velocity v is incident on a volume space of stationary filled with target particles having density n_g. The number of incident particles per unit volume at dx interacting with the target particles is dn.

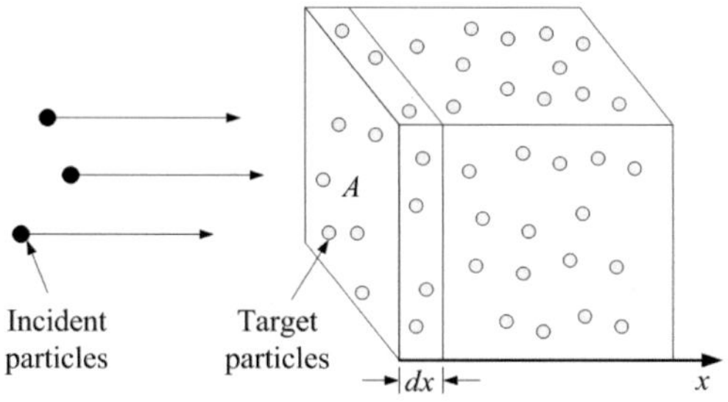

Figure 4.3: A flux of incident particles entering a space filled with target particles.

Then, dn is proportional to n, n_g, and dx for collisions within dx. Hence dn is written as

$$dn = -\sigma n n_g dx \qquad (4.1.1)$$

where the constant of proportionality σ has units of area and is called the cross section for the interaction. The minus sign denotes that the incident beam is removed from the initial flux. The interaction, e.g., ionization, excitation or scattering, must be specified in order to define a cross section. By multiplying Eq. (4. 1. 1) with v, a differential flux in dx is obtained as

$$d\Gamma = -\sigma \Gamma n_g dx \qquad (4.1.2)$$

In order to interpret σ, the incident and target particles are set as spheres of radii a_1 and a_2, and the interaction is a collision between the spheres. By collision, the circle of radius becomes $a_{12} = a_1 + a_2$ in the x-plane, and such interactions like scattering or energy transfer occur if the centers of the incident and target particles fall within this radius. Therefore, the fraction of the unit area for which a collision occurs is $n_g dx \pi a^2_{12}$, and the fraction of incident particles that collide within dx is then

$$\frac{d\Gamma}{\Gamma} = \frac{dn}{n} - n_g \sigma dx \qquad (4.1.3)$$

where

$$\sigma = \pi a_{12}^2 \qquad (4.1.4)$$

is the cross section of the spheres and is independent of v because the target particle is stationary. For electron collisions with atoms, the electron radius is negligible compared to the atomic radius so that $a_{12} \cong a$, the atomic radius. Eq. (4. 1. 2) is integrated to give the uncollided flux

$$\Gamma(x) = \Gamma_0 e^{-n_g \sigma x} \equiv \Gamma_0 e^{-x/\lambda} \tag{4. 1. 5}$$

The quantity

$$\lambda = \frac{1}{n_g \sigma} \tag{4. 1. 6}$$

is the mean free path for the decay of the beam. The mean time between interactions is then

$$\tau = \frac{\lambda}{v} \tag{4. 1. 7}$$

and its inverse is the collision frequency.

Collision is in general classified into elastic and inelastic collision. The total momentum and energy of the colliding particles after collision are equal so that the energy is conserved. When ions collide with atoms, the main processes are elastic scattering in which momentum and energy are exchanged. When electrons collide with atoms, only the internal energy of atoms is changed via excitation or ionization, maintaining the kinetic energy of atoms. The difference between the two collision processes is whether internal energy is conserved or not. To describe the collision in detail, a center of mass (CM) coordinate system is introduced as depicted in Figure 4.4. Incoming particles and target particles are treated equally at first, and the target particle is set stationary before the collision. Hence, a general collision is considered between two particles having mass m_i and m_t, velocity v_i and v_t, and scattering angle θ. The force acts along the line joining the centers of the particles. The velocity of the incident particle and the target particle along the CM line after collision is denoted as v_i' and v_t respectively.

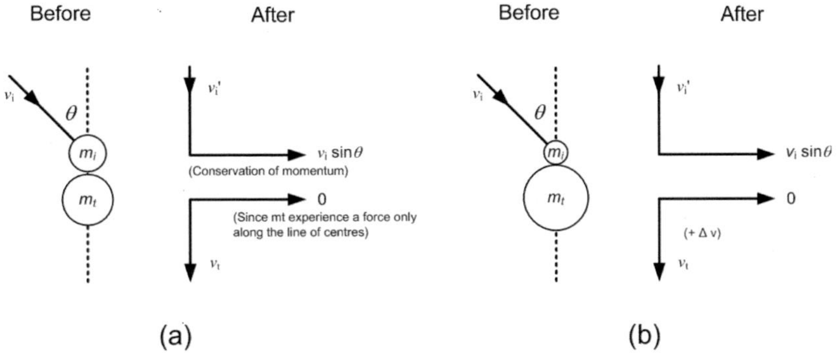

Figure 4.4: Energy components in (a) an elastic collision and in (b) inelastic collision, reprinted from Ref. [110].

(1) Elastic collisions

All kinetic energy is conserved in elastic collision processes. A case in which an incident particle is arriving to a target particle with a speed v_i is considered. The conservation of energy is written in the CM system as

$$m_i v_i \cos\theta = m_i v_i' + m_t v_t \quad (4.1.8)$$

and the total kinetic energy is also conserved, so that

$$\frac{1}{2} m_i v_i^2 = \frac{1}{2} m_i (v_i'^2 + v_i^2 \sin^2\theta) + \frac{1}{2} m_t v_t^2 \quad (4.1.9)$$

v_i' is eliminated by solving Eq. (4.1.9) and (4.1.10) to obtain

$$\frac{1}{2} m_t v_t^2 = \frac{1}{2} m_i v_i^2 \frac{4 m_i m_t}{(m_i + m_t)^2} \cos^2\theta \quad (4.1.10)$$

The fraction of energy lost by the incident particle is obtained as

$$\zeta_L = \frac{E_t}{E_i} = \frac{\frac{1}{2} m_t v_t^2}{\frac{1}{2} m_i v_i^2} = \frac{4 m_i m_t}{(m_i + m_t)^2} \cos^2\theta \quad (4.1.11)$$

The transferred energy becomes maximum when m_i is equal to m_t, and θ=0°. In Eq. (4. 1. 11) the first part, $4m_i m_t/(m_i+m_t)^2$ is known as energy transfer function. For hard-sphere scattering of electrons against atoms, electrons transfer little energy by elastic collisions with heavy particles because $m_e \ll m_i$, and the energy transfer function ζ_L is $4m_e/m_i \sim 10^{-4}$, allowing $T_e \gg T_i$ in a typical discharge. Hence a significant kinetic energy transfer occurs when the mass of two interacting particles is comparable.

However, the main collision processes in weakly ionized plasma occur between charged particles as well as uncharged neutral particles. It means that the collision process also includes coulombic interactions between electron-electron, electron-ion, or ion-ion. When charged and neutral particles collide in a weakly ionized plasma, the dominant process is short-range polarization scattering. It occurs for electrons at low energy and for ions scattering against neutrals, but at higher energies of electrons the collision time is shorter, therefore, the atoms do not have time to polarize. In this case, an inelastic process such as ionization or excitation becomes important.

(2) Inelastic collisions

Electrons colliding with atoms transfer their momentum and energy by elastic scattering, but they can change the internal energy of atoms in inelastic processes, resulting in excitation and ionization. Excitation of molecular vibrations and rotations are also important processes in molecular gases to describe the spectroscopy of atoms in a plasma discharge.

In quantum mechanics, the state of each electron in an atom is specified in terms of four quantum numbers, n, l, m_l and m_s (n, l, and m_l are integers), with the restrictions l

$+ 1 \leq n$, $|m_l| \leq l$, and with $m_s = \pm½$. The quantum numbers l and m_l specify the total orbital angular momentum and its component in a particular direction, and m_s specifies the direction of the electron spin. According to the Pauli exclusion principle no two electrons can occupy the same state. Hence stable atoms are built by placing electrons into the available states in order to increase energy. For example, the electronic configuration of argon is $1s^2 2s^2 2p^6 3s^2 3p^6$. The valence electrons are in the last subshell and determine the collision and other behaviors of atoms. An electron collision with an argon atom excites the atom to a higher energy level,

$$e + Ar \rightarrow Ar^* + e$$

corresponding to a change of state

$$3p^6 \rightarrow 3p^5 4s^1$$

for the valence electrons.

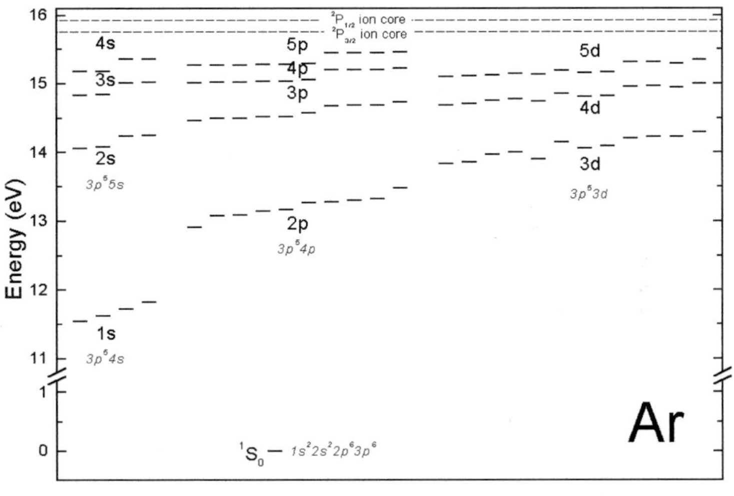

Figure 4.5: The energy levels of the argon atom, taken from Atomic collision group at Wisconsin university [111].

Atomic energy levels of Ar are indicated in Figure 4.5 with the dependence of the energy levels on the quantum numbers. If an atom is excited by collision or radiation, the bound state falls to lower energy state or the ground state emitting a photon by electric dipole radiation.

$$e + Ar \rightarrow Ar^* + e \rightarrow Ar + e + h\nu$$

For an excitation of an atom to a given energy level, energy under the ionization energy E_{iz} should be taken into account. Radiative transitions occur in form of electric dipole, electric quadrupole, or magnetic dipole radiation, but nonradiative transitions also occur to states of energy that subsequently radiate. These transition times are long compared to the collision time τ_e and τ_i. The energy levels of which electric dipole radiation is forbidden are called metastable. As an example, Ar has two metastable levels in the four levels of the $3p^54s$ configuration, the J=0 $1s_3$ and the J=2 $1s_5$. They are both metastable and are indicated as blue characters in Figure 4.5.

The radiation is usually in the visible and ultraviolet. Transitions from the energy levels allow a wide emission spectrum in the discharge. If the electron configuration of the atom is more complex, the spectrum becomes also complex. In a mixture of elements, each element will contribute its own set of distinctive lines, thus enabling one to establish which elements are present in the mixture. In Figure 4.6, the emission spectra of Ar from plasma are shown in separate graphs. Figure 4.6 (a) depicts the spectrum in vacuum UV range. Prominent emissions at the wavelength of 104.82 and 106.67 nm are found, which correspond to the transitions from the energy levels, J=1 $1s_2$ and $1s_4$, to ground state. Figure 4.6 (b) shows spectra in visible range, and various transitions from excited states are observable.

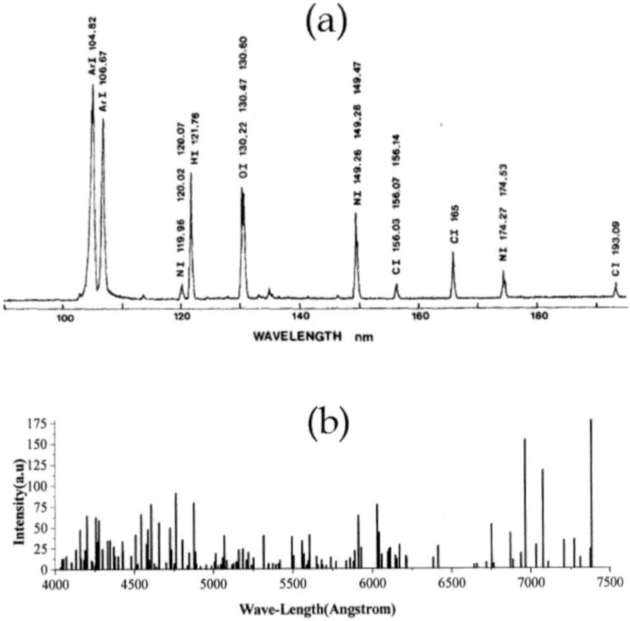

Figure 4.6: Emission spectra of Ar plasma scanned in the (a) UV and the (b) visible range, reprinted from references [112, 113], respectively.

4.2 Principle of magnetron sputtering

Magnetron sputtering is used in various thin film deposition applications. This is because sputtered films have properties which are often superior to those produced by evaporation sources due to the higher kinetic energies of sputtered and other particles generated during the sputtering. This has a favorable influence on the film adhesion and microstructure. Furthermore, sputtering methods meet the requirements of deposition uniformity and nonplanar substrates by enlarging the source target, and even insulating materials such as oxides, nitrides, and ceramics can be sputtered by applying a radio frequency (RF) source.

The fundamental operation of magnetron sputtering begins with a direct current (DC) glow discharge using a metallic material as a sputtering source and a cathode at the same time. The DC glow discharge has long been used as sputtering source for metallic materials, e.g., Al, W and Ti. A schematic illustration for a planar discharge in Ar gas driven by a constant DC source is demonstrated in Figure 4.7 (a). The upper cathode is assumed as Al, which acts as the target to provide Al atoms through a collision cascade from an ion impact. The energy of ions is acquired from a discharge voltage, and they are accelerated at the beginning of the plasma sheath. The ions are actually directed towards the target from a wide range before the sheath, and this region is called the presheath described as the Bohm sheath criterion. This ion impact generates other charge particles which are the main mechanism to sustain the plasma discharge. When an excess energy is transferred to a second electron in the solid, it is released from the solid and becomes free to move away from the solid surface. This process is called Auger emission, and the generated electron is called secondary electron.

The target-to-substrate (T-S) distance is typically 5~7 cm, and a high ion current density is required in order to achieve a sufficient deposition rate for industrial coating application. Therefore the discharge is operated in an abnormal glow regime with a high discharge voltage of 2 ~ 5 kV. In principle, high voltage can yield a high sputter yield Y, defined as the number of atoms sputtered per incident ion, which is valid with 500 ~ 1000 V ion energy. If the energy is increased beyond 1 kV, an ion implantation into the target occurs significantly so that the sputter yield decreases consequently. Also, a high operation pressure over 30 mTorr is required to maintain the discharge, and it leads to a considerable collision of sputtered atoms with Ar atoms. This effect results in a backscattering of the sputtered atoms so that the deposition rate is decreased and the film property becomes poor. Therefore it is desirable to operate the discharge at higher current density, lower voltages, and lower pressures.

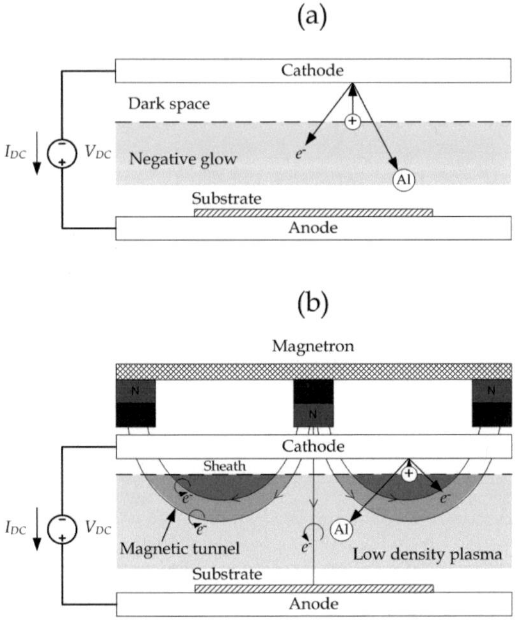

Figure 4.7: Schematic illustrations of (a) a DC glow discharge sputtering and (b) a DC magnetron sputtering.

By employing a set of permanent magnets at the back of the target cathode, magnetic field lines are generated as shown in Figure 4.7 (b). When a discharge is formed, the magnetic field confines the secondary electrons generated from the target surface. Thus high density plasma is maintained within the magnetic field, and it also leads to an increase of current density and an effective decrease of the operation pressure down to 2 ~ 5 mTorr.

The motion of charges acted by electric and magnetic fields are affected by the Lorentz force of

$$\mathbf{F} = q(\mathbf{E} + \mathbf{v} \times \mathbf{B}) \quad (4.2.1)$$

where **F** is the force, **E** is the electric field, **v** is the instantaneous velocity of the particle and **B** is the magnetic field. In uniform **E** and **B** fields, the charges move according to the Lorenz force, which is perpendicular to the directions of both fields. Since the ions are too heavy to react to the weak B field (~ 400 G), only the electrons are accelerated. The motion of the electrons is guided along a center with a gyration, the so-called **E** × **B** drift.

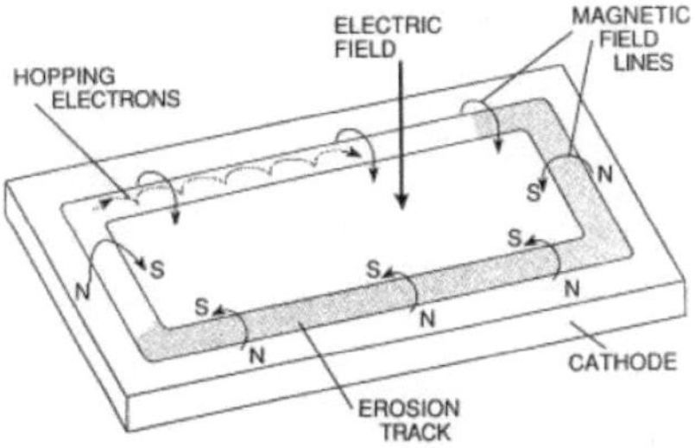

Figure 4.8: Applied electric and magnetic fields and electron motion in the planar magnetron sputtering, reprinted from Ref. [114].

The motion of the electron under uniform electric and magnetic field is illustrated in Figure 4.8. The continuous magnetic field lines are called 'magnetic tunnel', and the target surface within the magnetic tunnel is sputtered by ions. As illustrated, the electrons are accelerated by the Lorentz force along the tunnel, ionizing other neutral gases.

The discharge current at the dark space (cathode sheath) becomes different in the unmagnetized (DC glow discharge) and magnetized plasma sputtering. In the unmagnetized discharge at a high operating pressure such as 100 mTorr, the current at the edge of the cathode sheath region is carried only by the ions, so the carrier density can be approximated by

$$J_i = e n_s u_B \tag{4. 2. 2}$$

where n_s is the ion density at the sheath and u_B is the thermal velocity in the Bohm criterion $u_B=(eT_e/m_i)^{1/2}$. However, in the magnetized plasma where the operating pressure is low and the charge density is high, the collisionless Child law can be used to describe the flow of the ions to the cathode as

$$J_i = \frac{4}{9}\varepsilon_0\left(\frac{2e}{m_i}\right)^{1/2}\frac{V_{dc}^{3/2}}{s^2} \tag{4. 2. 3}$$

where V_{dc} is the cathode voltage and s the sheath thickness. This is also the well-known form of SCLC in a planar diode.

The current components in the plasma can also be altered by a power source and corresponding operation mode. As discussed above, the plasma with DC power is sustained by the secondary electrons generated at the target surface. The electrons collide with neutral gases so that charged particles are generated in the bulk plasma, and the conduction in the bulk plasma is carried out through the ionized medium.

However, when a time-varying electric field is applied to the cathode, the free electrons begin to oscillate. This results in a displacement of free electrons to ions in the plasma. If the frequency of the electric field reaches the microwave region (1-10 GHz), the electrons oscillate at the natural frequency

$$\omega_{pe} = \left(\frac{e^2 n_0}{\varepsilon_0 m} \right)^{1/2} \tag{4. 2. 4}$$

which is the electron plasma frequency. The ions do not respond to the time-varying electric field because they have very low plasma frequency compared with that of the electrons. Here a frequency of 13.56 MHz is taken into account, which is the most widely used frequency in RF region for plasma processes.

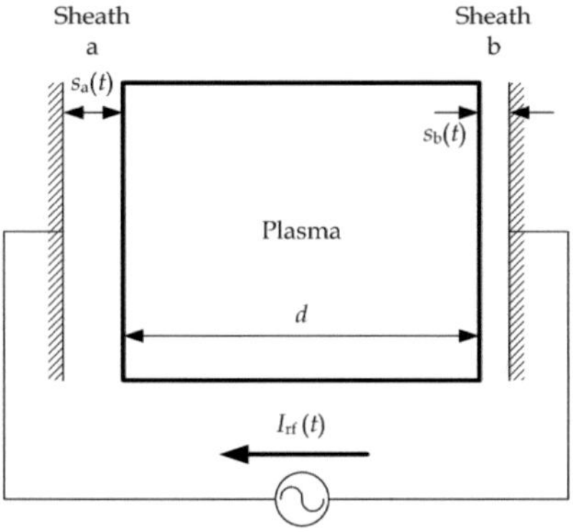

Figure 4.9: Basic RF discharge model with time-varying sheath. A displacement of the plasma is illustrated.

In Figure 4.9, a basic RF discharge model in a unmagnetized plasma is shown. Between the two parallel plates, electrons are assumed to respond only to the time-averaged potential. This is because the ion plasma frequency is much lower than the natural plasma frequency due to their heavy mass. Only electrons respond instantaneously to the potential, and they carry the RF discharge current.

The total current flowing across a one-dimensional discharge is given as

$$J_T = \varepsilon_0 \frac{dE}{dt} + J_{cond} \tag{4. 2. 5}$$

where the first term in the right hand side is the displacement current in the plasma sheath and J_{cond} is the conduction current density by the motion of the free charges in the bulk plasma. There are two main mechanisms for heating electrons in RF plasma: ohmic and stochastic heating.

The sheath boundary a and b oscillate linearly with the displacement current. In this case, the electrons are reflected and accelerated with a moving high-voltage sheath according to the varying electric field. This results in a momentum transfer from the high voltage leading to stochastic heating. In a simply approximated form, the stochastic electron power deposited for a single sheath is given with the known current density J as

$$S_{stoc} = \frac{1}{2} \frac{m\upsilon_e}{e^2 n} J \qquad (4.\ 2.\ 6)$$

where υ_e is the mean electron speed. In the bulk plasma, the oscillating free electrons collide with the neutrals and transfer momentum. Therefore, ohmic heating mechanism dominates in the bulk plasma, and the power per unit area transferred by the ohmic heating in the bulk plasma with the length d is obtained as

$$S_{ohm} = \frac{1}{2} J^2 \frac{d}{\sigma_{dc}} \qquad (4.\ 2.\ 7)$$

where J is the current density in the bulk plasma and σ_{dc} is the DC plasma conductivity. In the bulk plasma, a cold plasma approximation is applied for low frequency, and a DC plasma conductivity σ_{DC} is found to be

$$\sigma_{DC} = \frac{e^2 n_0}{m \nu_m} \qquad (4.\ 2.\ 8)$$

where ν_m is the electron-neutral collision frequency. Substituting Eq. (4. 2. 8) into Eq. (4. 2. 7), S_{ohm} is obtained

$$S_{ohm} = \frac{1}{2} J^2 \frac{m \nu_m d}{e^2 n} \qquad (4.\ 2.\ 9)$$

where ν_m is the mean collision frequency. The total time-average electron power per unit area is then obtained by adding Eq. (4. 2. 9) and two sheaths of Eq. (4. 2. 6)

$$S_e = \frac{1}{2} \frac{m}{e^2 n} (\nu_m d + 2\upsilon_e) J^2 \qquad (4.\ 2.\ 10)$$

Due to the additional stochastic electron heating mechanism, the charge density and the electron temperature in a RF plasma are much higher than in a DC plasma.

When a magnet field is introduced, a number of effects follow that increase the plasma density and reduce the sheath voltage. This is because the magnetic field can confine the energetic electrons to a small volume near the electrode. Thus a transverse magnetic field can increase the efficiency of the stochastic heating due to multiple collisions of electrons with the oscillating sheaths, resulting in the increase of the plasma density. Accordingly, there is also an increase in the efficiency of the ohmic heating due to higher electric fields in the bulk plasma.

The sputtering mechanism in the RF plasma should be also discussed. As discussed above, the ions can not respond to the high frequency. In the RF plasma magnetron sputtering, however, the surface of a smaller electrode is charged negatively by excess electrons due to the difference of the electrode areas. Normally the sputter cathode is much smaller than the grounded chamber so that an accumulation of the electrons occurs on the target. The accumulated electrons then induce a negative DC bias voltage so that the ions are accelerated towards the target and sputter it. This bias is called 'self-bias'. Since the self-bias is as low as several tens volts in normal process regime, the deposition rate using the RF power is quite low. In order to improve the deposition rate, a small amount of DC power can be superimposed into the RF power. This technique will be introduced in section 4.3.2.

The properties of thin film deposited by magnetron sputtering and the energy of the generated particles are significantly influenced by process parameters, power sources and hardware configurations. For the sputter deposition of metals, DC power is usually used with Ar process gas. The Thornton model is widely referred to for the sputter-deposited film structure and topography [115].

Discharge power for magnetron sputtering deposition

Sputtering rate: Discharge power P is given by the product of discharge voltage V and current I, namely $P = V \times I$. The discharge voltage determines a sputtering yield Y in a given projecting/bombarded system by the acceleration energy of ions bombarding a target surface. The discharge current usually varies at a larger rate than the voltage with respect to the power in a normal magnetron sputtering. As discussed earlier, the current in DC discharge represents the number of incident ions, namely, ion current flowing through the target cathode. Hence it can also reflect the number of sputtered atoms, consequently, deposition rate.

Film properties: Discharge power-dependant deposition rate affects the nucleation rate on the substrate. If the flux of vapor, i.e. high deposition rate, is high, then the nucleation rate increases. The high flux results in a high concentration of nuclei on the surface and, consequently, a small grain size. Inversely, low flux leads to a low concentration resulting in a larger grain size.

Process pressure for magnetron sputtering deposition

Transport: The process gas is an essential element in plasma discharge. A low pressure gas environment in which sputtered particles are transported from the target to the substrate provides less gas phase collisions, but such a gas phase collision and a resulting thermalization of energetic particles occur more frequently at a higher gas pressure. This effect appears normally as a lower deposition rate.

Film properties: In many sputtering depositions, the process pressure becomes a key parameter by which morphology or intrinsic stress of film can be controlled. In general, the variation of process pressure is helpful to achieve a low film stress of films like W, Cr or TCOs that tend to show high film stresses after sputtering deposition. These effects are all associated with a bombardment of the energetic particles. By higher process pressure, the bombardment can be reduced due to the collision and the thermalization of atomic particles.

Since the sputtering is based on the momentum transfer from the ions with high incident energy (300~600 eV) from the plasma, the energy transfer process generates not only the secondary electron, but also various kinds of atomic particles having high kinetic energy.

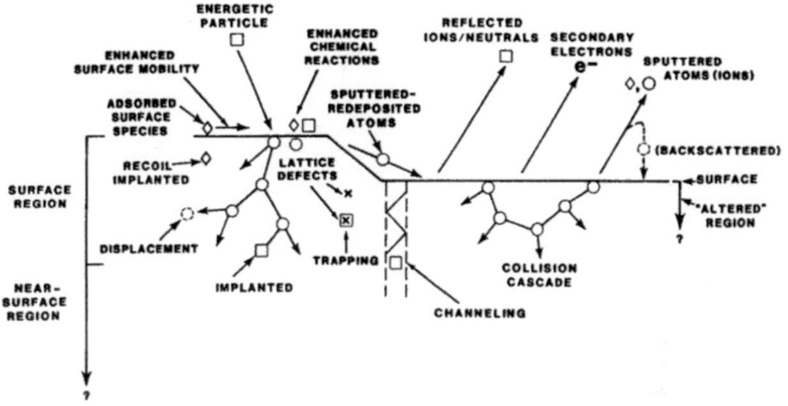

Figure 4.10: Events that occur on a surface being bombarded with energetic atomic-sized particles. Reprinted from Ref. [116].

Figure 4.10 shows the processes that can occur on the surface and in the near-surface region of the bombarded surface with highly energetic particles. Most of the transferred energy (>95%) dissipate as heat in those regions, but some of the bombarded particles are reflected as high energy neutrals. The generation of the recoiled particles follows the momentum transfer of Eq. (4. 1. 11). Depending on the mass relation of the particles, the neutrals can be generated when the mass of the incident particle is lighter than that of the bombarded surface atom. Therefore, the energy of the recoiled particle is proportional to the discharge voltage. Since the discharge voltage is greatly affected by a magnetron design and a hardware configuration, the resulting effects can be quite different depending on the sputter equipment. The quantity of the recoiled particles is determined by the number of the incident particles, the ions in sputtering, thus the influence of the discharge current is discussed in detail here. Rossnagel reported an effect of energetic particle bombardment on a substrate and the behavior with respect to process parameters [117]. According to the study, the local gas density near a magnetron cathode undergoes large changes during the sputtering process due to the heating and rarefaction of the gas by collisions with energetic atoms including recoiled neutrals and sputtered atoms. If the amount of incident energy (flux of sputtered atoms times their average kinetic energy), the gas density, and the thermal conductivity of the background gas are given, a limiting form of the local gas density near the cathode is expressed as

$$n_g = \left(\frac{2 n_0 T_0 \pi f K}{E_a Y \sigma} \right)^{1/2} I^{-0.5} \tag{4. 2. 11}$$

where n_0 is the original gas density, T_0 the wall temperature, K the thermal conductivity of the gas, E_a the average kinetic energy of the sputtered atoms, Y the sputter yield, σ the cross section for momentum transfer, and f the number of mean free paths away from the cathode before the particle is thermalized. As seen in Eq. (4. 2. 11), the local gas density is an inverse square root function of discharge current I. The decrease of n_g is because the energetic particles like Ar neutrals and sputtered atoms collide with the local gases eliminating them from the near-cathode region. In other words, more energetic particles are produced with increasing the discharge current.

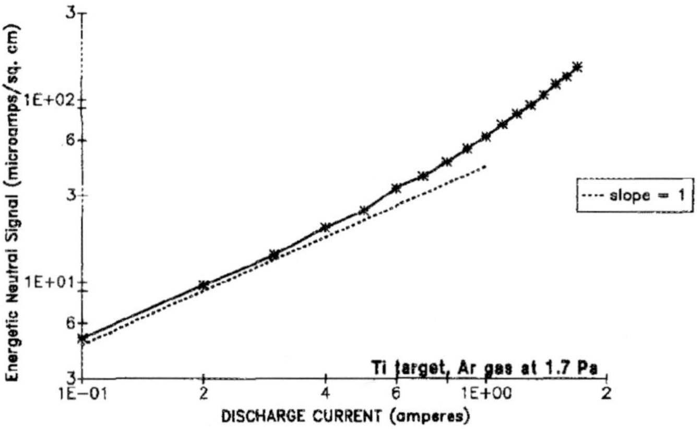

Figure 4.11: Energetic neutral flux (in mA/cm^2 equivalent units) as a function of discharge current for a Ti cathode in Ar at 12 mTorr. Reprinted from Ref. [117].

Rossnagel clarified that effect by measuring the signal of Ar neutrals recoiled from the Ti target sputtering with Ar gas, and the result is shown in Figure 4.11 [117]. In the study, a 7.5 cm diameter circular sputtering target (44.16 cm^2) was used, so the current density at the test regime is derived as 11.04 mA/cm^2 calculated from the target area and the current of 4 A, which exhibits still a linearly increasing Ar neutral signal.

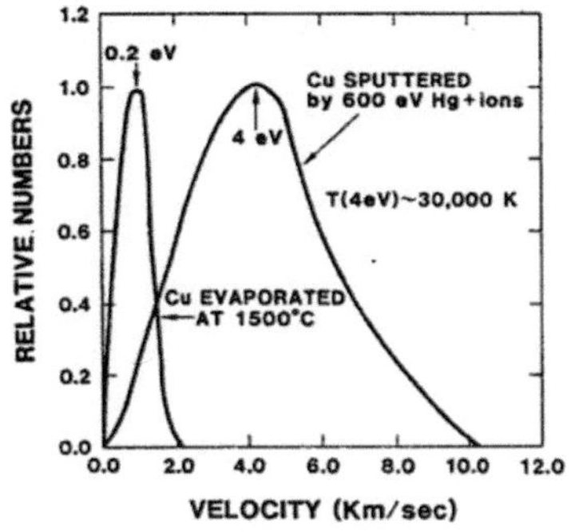

Figure 4.12: Calculated energy distribution of atoms leaving a sputtering target. The energy distribution is maxwellian showing a tail at high kinetic energy. Reprinted from Ref. [116].

The sputtered target atoms have also high kinetic energy. In Figure 4.12, energy distributions of evaporated and sputtered Cu atoms are shown. The energy of the sputtered atoms exhibits also a Maxwellian distribution like electrons in plasma. The average energy of sputtered atoms is just two or three eV. However, there exists a high energy region, so an effective impact can be given to the bombarded layers on the surface of substrate.

Pressure related phenomena in plasmas and deposition behavior in magnetron sputtering have been described by many researchers [118-120]. For an explanation of the phenomena, collision effects between particles during transport should be discussed concretely. In section 4.1 basic collision and energy transfer were briefly introduced. Since the energetic particles generated collide with the background gas atoms, the process pressure and the energy of the background gases should be taken into account into Eq. (4. 1. 6). The mean free path λ_m is then expressed as

$$\lambda_m = \frac{k_B T}{p\sigma} \tag{4. 2. 12}$$

where k_B is the Boltzmann constant and T the temperature of the sputtering gas. If the pressure increases in magnetron sputtering, the deposition rate is decreased due to collision between particles and a resulting backscattering effect. The energy of the particles is then thermalized by energy transfer to the process gases. The kinetic energy of the sputtered atoms is quite high over 10 eV. Drüsedau *et al.* investigated the thermalization and resulting variation in deposition rates [119]. The dependence of the deposition rate on pressure showed two contributions by directed and diffusive transports. A high deposition rate at low pressure was attributed to the directed transport with less collision, and a lower deposition rate at a high pressure was vice versa to the diffusive transport with backscattering. The average energy of the sputtered Al atoms was also calculated to reach 19.2 eV which is much higher than the sputtered Cu atoms in Figure 4.12. This energy is so high that the Al atoms can act significantly on the surface of substrate or underlying layers. It is thus important to reduce the kinetic energy dramatically in order to lower energy transfer.

Summarizing, the total energy transfer from the discussed particles appears as a heat load onto the substrate. Several factors contribute to the substrate heating. These are the heat of condensation, the kinetic energy of the sputtered atoms, bombardment by energetic neutrals, plasma radiation, gas collision and heating, and the energy transfer by ions and electrons onto the substrate. Referring to the study of Epke *et al.*, the total energy transfer in magnetron sputtering is then summarized as

$$E_{total} = U + E_k + E_0 \frac{R_E}{Y} + E_p + E_i + E_e + E_t \qquad (4.\ 2.\ 13)$$

where U and E_k represent the heat of condensation and the average kinetic energy of the deposited atoms, respectively [121]. The third term represents the contribution of the reflected gas neutrals determined per sputtered atom in terms of the yield Y. E_0 is the incident energy of the ion and R_E is the reflection coefficient. E_p represents the contribution of plasma radiation, which is photonic radiation from excited atoms. E_i, E_e and E_t are the contributions of ion, electron and thermal radiation, respectively. Each term in Eq. (4. 2. 13) is significantly dependent on process parameters, so their energy variations must be investigated in order to understand the behavior of the particles generated.

The generation of energetic atomic particles should be discussed depending on the target material. As already explained, sputtering processes generate various kinds of energetic particles. The energetic atomic particles include recoiled Ar neutrals, ions, or sputtered particles. The generation of the neutral atoms in the magnetron sputtering is the result of the reflection and Auger neutralization of the incident Ar ions directed at the target surface. This neutral atom is, however, strongly dependent on the mass relation of the target material and the incident ion.

In case of the Al sputtering with Ar gas, when the atomic masses of Ar (39.95) and Al (26.98) are applied to Eq. (4. 1. 11) with an incident angle of 0°, the energy transfer function $4m_im_t/(m_i+m_t)^2$ becomes 0.962. Hence a nearly perfect energy transfer is accomplished, and the recoiled Ar neutral is hardly generated in the Al sputtering using Ar gas. Therefore, only the sputtered Al atoms remain as the energetic atomic particles in Al sputtering with Ar gas. The flux of the energetic Al atoms is presumably proportional to the discharge current alike the recoiled neutral that was examined by Rossnagel. The third term in Eq. (4. 2. 13) is then neglected, and only the energy term of the sputtered Al atoms should be taken into account.

TCO sputtering provides a more complicated situation. In ITO sputtering, for example, a small amount of oxygen gas is required to reduce the resistivity of the film. However, O_2^- ions generated during ITO sputtering are the source that affects the structural and electrical property of ITO film negatively [122-124]. They are accelerated by the negative voltage applied to sputter cathode so that their kinetic energy is in principle the same as the cathode voltage (100 ~ 200 eV). Their amount is very little compared to Ar sputtering gas, but enough to give a critical impact. Besides the O_2^- ions, recoiled Ar neutrals can also be the source for damage. Since the atomic masses of ITO-consisting materials, In (114.82) and Sn (118.71), are much heavier than Ar (39.95), the energy transfer function $\zeta_L=4m_im_t/(m_i+m_t)^2$ according to Eq. (4. 1. 11) is derived as 0.77 and 0.75 for In and Sn, respectively. This means that the Ar ions are recoiled with a rest energy of around 25% of incident energy. The recoiled Ar neutrals would spread in all directions by a surface scattering with the energy correspondent to the rest energy. From this point of view, a low voltage sputtering is needed, and RF superimposed DC power or a strong magnet field can be appropriate solutions to control the energy of the particles for TCO sputter deposition.

4.3 Power supply systems for sputtering

4.3.1 DC and RF power

The most prominent advantage of magnetron sputtering is that any solid material can be deposited on various substrate types. There are various power supply systems applicable for magnetron sputtering deposition, depending on the target materials. Metal sputtering is usually performed by supplying DC power because the target cathode is conductive. Insulating materials, such as SiO_2, Al_2O_3 or AlN, are however not able to be sputtered by DC power because the insulating targets do not allow current flow. These insulating materials need therefore an other discharge power system, for example, RF power supply at high frequency over 50 kHz in order to transfer the power through the target.

The mechanism of the RF power sputtering is explained by the self-bias that appears on the target surface. RF frequencies used for sputter deposition lie in the range of 0.5-30 MHz, and 13.56 MHz is the frequency commercially used for plasma processing. RF power sputtering has however a disadvantage since most insulating materials have poor thermal conductivity, high coefficients of thermal expansion, and are usually brittle. As discussed already, most of the bombarding energy is converted into heat such that the large thermal gradients can be generated in the target, which can result in a fracturing of the target at high power or long process time. Since the sputtering in RF plasma is carried out by ions with the energy correspondent to the self-bias, the sputter yield is usually very low. Therefore the high power and the long process time are unavoidable in RF power sputtering for a film with certain thick thickness.

4.3.2 RF super-imposed DC power

RF super-imposed DC power (RF+DC) power has the advantage that RF and DC power are well combined in film deposition. As discussed above and in section 4.2, the RF power sputtering results in a low discharge voltage and leads to an effective ion bombardment on the substrate due to plasma oscillation and large difference between plasma and floating potential V_p-V_f [125, 126]. The ion bombardment with a moderate energy can assist the film growth and lead to better and denser films.

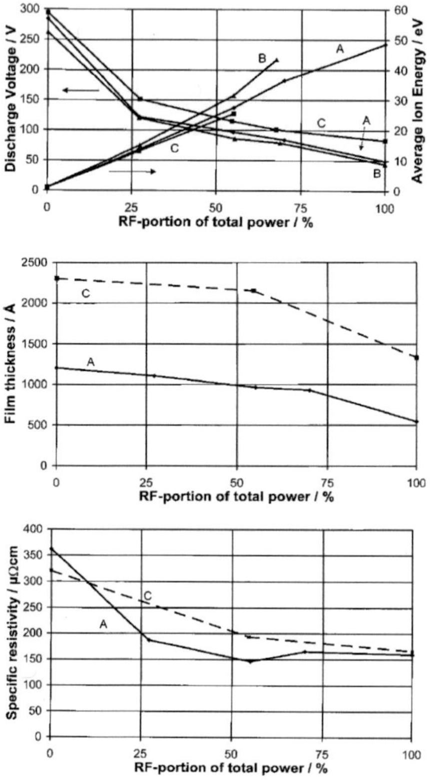

Figure 4.13: Behavior of discharge voltage V, sputter rate, and film resistivity in an RF superimposed DC sputtering process of ITO as a function of RF-portion. A, B and C denote process parameters: A is total power of 1000 W and process pressure of 0.2 Pa. B is power of 1,000 W and process pressure of 0.4 Pa. C is power of 2,000 W and process pressure of 0.2 Pa. Reprinted from Ref. [127].

However, the lower discharge voltage results in a lower deposition rate and, accordingly, equipment cost increases in order to apply high power for a moderate deposition rate on large area substrate. By adding a certain amount of DC power to the RF power, it is possible to obtain a reasonably high sputter rate with the substrate ion bombardment. In Figure 4.13, results from ITO deposition using RF + DC plasma observed by Bender *et al.* are shown [127]. An optimal power ratio is found at the RF portion of 50% satisfying both deposition rate and ITO film resistivity. Fast changes of discharge voltage, deposition rate, and resistivity between the RF portion of 25 and 50% are due to a transition of plasma excitation. In order to explain the transition, the waveforms of both powers must be examined.

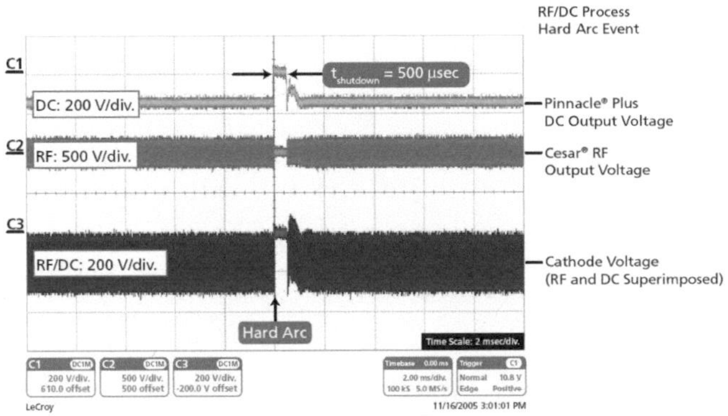

Figure 4.14: Pinnacle Plus power supply output voltage (C1), RF generator output voltage (C2), and cathode voltage (C3) in a RF superimposed DC sputtering process at the RF portion of 33.33%. Reprinted from Ref. [128].

Figure 4.14 shows the oscillogram of voltage outputs from a RF+DC power experiment. Voltages from three channels are exhibited: C1 is the output voltage from DC power supply (Pinnacle® Plus, advanced energy) at the DC power of 500 W, C2 is from RF power supply (Cesar®, advanced energy) at the RF power of 250 W, and C3 is the cathode voltage of the RF+DC. The portion of RF with respect to the total power is 33.33 %. The DC discharge bias indicates approximately constant 150 V, and this value is affected by process condition and cathode design such as magnetron strength. 150 V is a rather low discharge voltage which can be obtained in normal magnetron sputtering. The RF discharge voltage alters within ± 200 V which is larger than the DC discharge bias. However, this high RF discharge voltage, as discussed in the RF plasma part, can not attract ions because they are too heavy to follow the quickly alternating voltage. Sputtering from ion bombardment occurs only by self-bias. When both powers are superimposed, the voltage of the RF+DC becomes C3, where the discharge voltage alters from 0 V to 300 V. In this regime, the plasma maintains still a negative DC discharge, namely DC mode. However, the discharge voltage starts to alter for both polarities if the RF portion is further increased. Then electrons begin to oscillate between the cathode target and the substrate, and the plasma area gets broadened near to the substrate due to further ionization by the oscillating electrons. The energy of ions also increases, and thus the ion bombardment is enhanced, falling onto the substrate at several eV.

RF superimposed DC sputtering processes show favorable effects for the growth of various thin films and the plasma controllability. In particular, the behavior of the discharge voltage varies dramatically with the mixing ratios of RF and DC power and process pressures. The variation of the discharge voltage may help to change the energy of energetic particles produced during magnetron sputtering. However, it should be noted that only a conductive target can benefit from the RF superimposed DC power.

4.3.3 Pulsed DC power

As an alternative way to obtain an insulating or dielectric film, a metallic target can be sputtered with reactive gases, e.g. O_2 and N_2, and DC power to obtain metal oxides or nitrides films. However, serious arcs are generated due to reacted target surface and resulting charge accumulation. Also, an "anode disappearing" effect is found when sputtering is carried out with continuous DC power, resulting in plasma extinction eventually [129]. Furthermore, the surface coated with non-conductive layers accumulates electric charges from the plasma, and these charges create mini- or macro-arcs which result in a very uneven removal of material from the target electrode and particle formation. These particles are sometimes incorporated into the deposited films preventing their use in high-tech applications such as thin film devices, which require high purity of the thin films. In order to eliminate the charges on the target surface, a pulsed DC power is used generating a unipolar or bipolar square waveform which operates at 50-250 kHz [130, 131].

In case of the unipolar pulsed DC, the power is applied to the target for a time τ_{on}, the on-time during a negative voltage pulse. At the end of the on-time, the power is switched off for a period of time τ_{off}, the off-time. During the off-time the accumulated charges are discharged, but not perfectly. When a bipolar power is supplied, a small positive voltage (about 20 V) is commonly applied during the off-time. In this case, the off-time is sometimes referred to as the reverse time, τ_{rev}. The dielectric surfaces that are charged up during the on-time are more effectively discharged during the reverse time. Arc-free operation requires the on-time to be sufficiently short to avoid charge build-up that can cause breakdown and arcing. In contrast, the reverse time should be sufficiently long to entirely discharge the surfaces in order to avoid charge accumulation. In Figure 4.15, a typical voltage sequence used in the bipolar pulsed DC sputtering with a single magnetron is shown. This waveform is asymmetrical, but the ratio of the on/off-time can be adjustable.

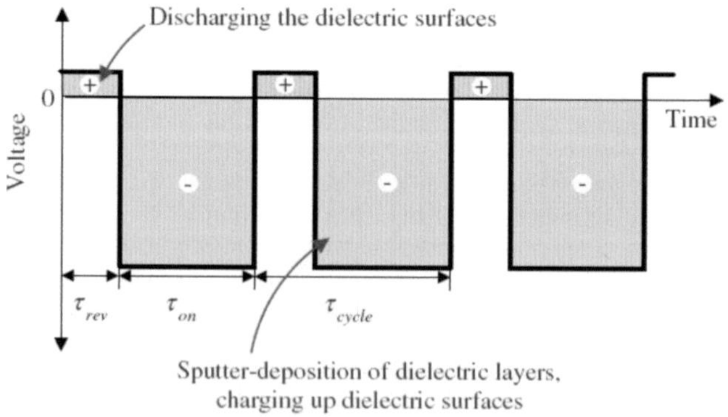

Figure 4.15: Ideal voltage sequence applied to asymmetrical bipolar pulsed DC sputtering of dielectrics. Reprinted from Ref. [132].

When equipped with two power supplies and two magnetron sputter cathodes, the dual magnetron can be realized. By implementing a bipolar switching unit, the input power is transformed to the unipolar or bipolar square wave pulses [133]. If the switching unit contains independent channels, two magnetron cathodes can be operated independently as well, enabling two different processes with different target materials. This method is called dual magnetron sputtering (DMS).

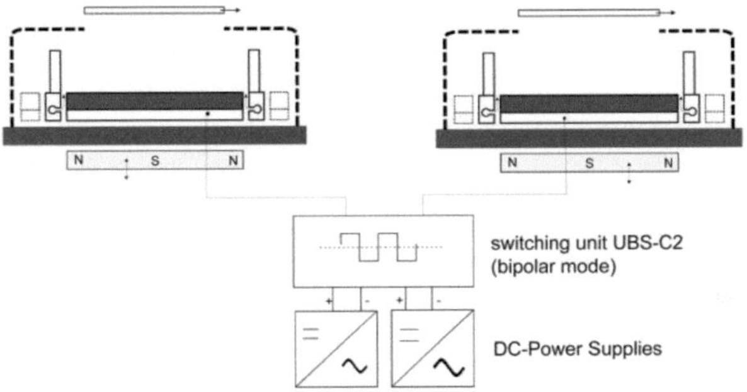

Figure 4.16: Schematic diagram of bipolar powering configuration of two magnetron sputter cathodes. Reprinted from Ref. [133].

In Figure 4.16, the bipolar pulse power system and hardware configuration with two magnetron sputter cathodes are shown. During one pulse cycle, both targets are biased positively and negatively, which act as anode and cathode, respectively. If the anode shield in front of the target is removed, a common discharge space can be formed. Many researchers have been working with this technique, but the Fraunhofer Institute for electron beam and plasma technology (FEP) has developed a unique pulse unit (UBS-C2) which fulfills all the requirements mentioned above. Two photographs of a plasma discharge of the dual magnetron sputtering are shown in Figure 4.17. Each of the magnetron cathodes is operated independently out of phase so that the plasma is also formed between the two targets, because one target acts as an anode grounded in unipolar pulse or biased positively in bipolar pulse.

Figure 4.17: Photographs of plasma discharge in a dual magnetron sputter system with unipolar pulsed power. Photographs are from Fraunhofer Institute for electron beam and plasma technology, Dresden, Germany.

4.4 Facing target sputtering (FTS)

Facing target sputtering (FTS) was developed for the sputtering of magnetic materials and reactive sputtering [134, 135]. In case of a sputter cathode with a magnetic target material, such as Ni, Co or Fe, the whole magnetic flux formed by the arrays of the permanent NdFeB magnets passes almost through the magnetic target material, so that magnetic tunnel on the target surface is hardly generated. This magnetic flux is very essential to confine secondary γ-electrons which are the main factor to sustain the plasma. On the other hand, facing target sputtering system can sputter even magnetic targets with the help of the magnetic fields formed between two parallel targets. The FTS cathodes basically consist of two targets and two magnetrons. The magnetrons are installed with opposite polarities in order to form magnetic flux between magnetrons. The secondary γ-electrons generated are effectively confined by the magnetic fields, and they can promote the ionization of the gas in the space. A schematic illustration of FTS is depicted in Figure 4.18.

As seen in Figure 4.18, the electrons are confined by the magnetic field, ionizing background Ar gases. The Ar^+ ions are then accelerated to the target surface by the electric field, and secondary γ-electrons are generated alike usual sputtering. Ar neutrals are also generated depending on the energy transfer function ζ_L.

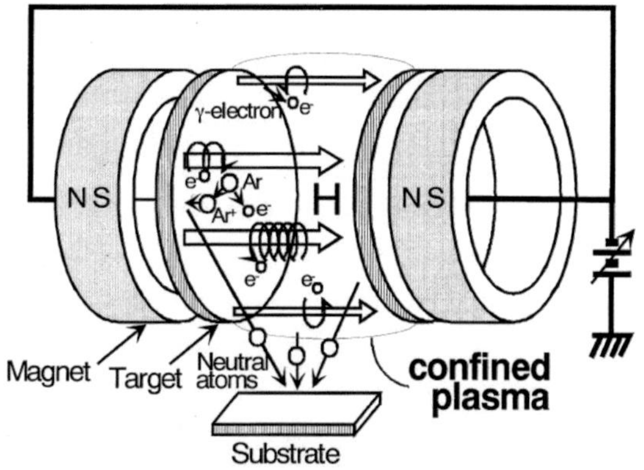

Figure 4.18: Schematic illustration of FTS system taken from the Ref. [135].

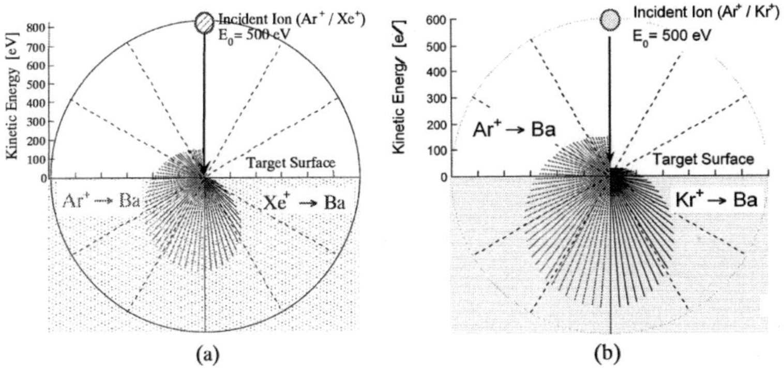

Figure 4.19: Angular distribution of kinetic energy of recoiled particles by Ba atoms on the target surface, reprinted from the reference [135]. Different gas mixtures are used (a) Ar+Xe, (b) Ar+Kr.

Figure 4.19 shows the calculated distribution of kinetic energy of reflected neutrals [135]. In Ba sputtering with Ar^+ ion, Ar^+ ions are undoubtedly reflected from Ba target because the atomic mass of Ar (39.95) is smaller than that of Ba (137.33). Therefore, the FTS system has still a disadvantage for such circumstances in which the target material is heavier than the process gas. However, FTS has a clear merit in avoiding an irradiation of electrons or ions from bulk plasma because the plasma is effectively confined by the magnetic field between two targets and the substrate is usually set away from the sputter cathode. This merit is especially beneficial to avoid the bombardment of negatively charged atomic particles in insulating film deposition with reactive gases like oxygen and nitrogen, which are ionized negatively and accelerated by the cathode voltage. Therefore, the FTS system represents a large advantage for TCO deposition in reducing the bombardment of such reactive species.

4.5 Degradation in OLED after sputtering on organic material

There are many studies regarding top metallic electrode deposition on organic layers by means of magnetron sputtering [50, 136], but critical damage was reported. Furthermore, the energetic particles were described without an exact identification and classification. The degradations were mostly explained that high energetic particles like recoiled Ar neutrals and ions from plasma are the dominant damage sources. Liao *et al.* evaluated the sputter damage with the effect of Ar^+ ion bombardment on tris-(8-hydroxyquinoline) aluminum (Alq_3) molecules, and they explained that the bombardment of high energetic particles destroys the molecular band structure, resulting in a metallic surface [5].

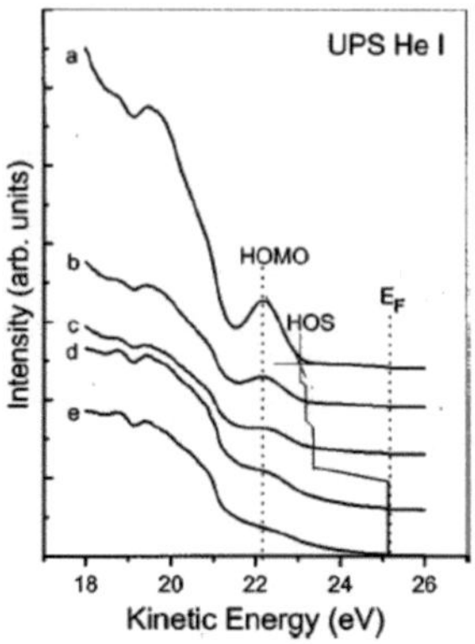

Figure 4.20: He I UPS spectra of Alq_3 film as a function of increasing Ar^+ (100 eV) irradiation dose. The irradiation doses of curves a, b, c, d, and e are 0, 6.0×10^{13}, 1.2×10^{14}, and 8.0×10^{14} ions/cm^2, respectively. Reprinted from Ref. [5].

In Figure 4.20, UPS spectra obtained from an Alq_3 thin film after Ar^+ ion irradiation at various ion doses are shown. Before the ion irradiation, the HOMO structure was clearly resolved, and the HOMO cutoff that was described as the highest occupied state (HOS) in that study was approximately 2.13 eV below the E_F. After the irradiation, the density of HOMO decreased and the HOMO cutoff extended towards the Fermi edge. When the irradiation dose reached 2.6×10^{14} ions/cm^2, the HOMO cutoff moved to the E_F, indicating that the damaged surface became metallic. The authors expected that the LUMO is also degraded and extended downwards. So a postulate that the HOMO and the LUMO are eventually overlapped after the bombardment from the highly energetic particle was introduced as depicted in Figure 4.21. This metallic surface was thus presumed to modify injection from the electrode and transport of charge carriers, and to result in a large leakage current.

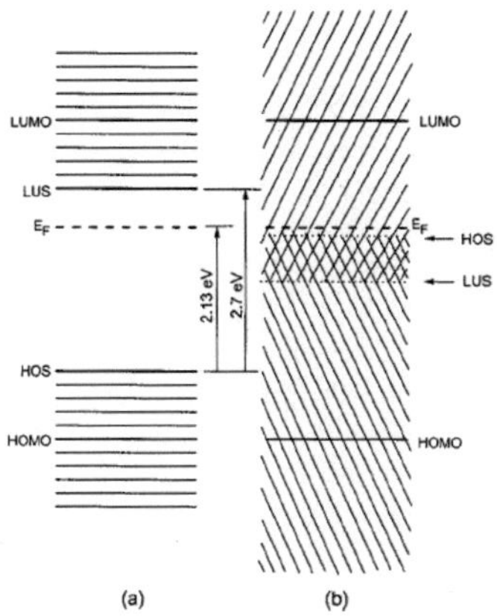

Figure 4.21: Schematic diagram of energy level of the Alq_3 film: (a) as-deposited and (b) irradiated by 100 eV Ar^+ with a dose of 8×10^{14} ions/cm^2. Reprinted from Ref. [5].

This explanation provides a major reason responsible for a large leakage current through the OLED. However, the experiments were not performed by a substantial sputtering deposition, and the explanation was only limited to an ion bombardment effect and resulting molecule destruction. As discussed earlier in this study, various kinds of energetic particle as well as ion bombardments exist in real sputtering process. They are sputtered target atoms, secondary electrons and radiations including vacuum ultraviolet-ultraviolet (VUV-UV) photons. Their energies are significantly exceeding the binding energy of carbon-carbon covalent bonding or carbon-hydrogen bonding, typical for organic molecules.

Although there were many studies regarding the degradation by the energetic atomic particles, radiation-induced degradation was scarcely discussed. VUV induced damage has been studied widely in MOS technology, bipolar transistor, or surface modification engineering of polymers using plasma [137, 139]. High-energy photons ($h\nu > 9$ eV) are generated from recombination and relaxation processes in the plasma. Depending on the energy of the incident photons, two cases that may introduce damage can occur in MOS: (1) photons with energies higher than the band gap of SiO_2 (9 eV) generate electron–hole pairs in the oxide, and (2) photons with energies lower than the SiO_2 energy gap but greater than 4.2 eV (the height of the minimum energy barrier between the Si substrate valence band and the oxide conduction band) cause electron injection from the silicon surface into the oxide through the photoelectric effect. In case (1), it has been established that electron–hole-pair generation increases the bulk and interface trapped-charge density, which will affect device reliability accordingly. In case (2), there is a controversy: while some studies show a similar phenomenon to the one proposed, others report that photocarrier injection from the substrate is beneficial, proposing that it anneals the positive interface charge through a recombination process. In section 6, it will be discussed whether the first mechanism is suitable for metal/organic interface or the second.

The energy transfer from each particle is affected differently from the kind of discharge gas, target material, process parameter, and power source. Since those particles are generated simultaneously, having energy disparity or same polarities like ionized target atoms and Ar ions in cathode sheath and bulk plasma, the identification of the dominant damage source by means of plasma measurements has been thus hindered in understanding the exact damage mechanisms up to now.

Not only metallic electrode, various electrode materials, and power sources have been attempted by the sputter deposition. There were several trials to deposit ITO films on OLED in order to realize a highly transparent electrode on Alq_3 based organic layers. Due to low glass transition temperature of organic materials, ITO should be deposited at a low temperature near room temperature. Magnetron sputtering was widely used in the studies. Gu *et al.* reported a transparent organic light emitting device [140]. ITO was deposited on thin (50-400 Å) Mg-Ag layer by RF magnetron sputtering in order to increase the conductivity of the cathode without loosing transparency. The role of the Mg-Ag cathode was not only a conductive electrode, but also a buffer layer to protect underlying organic layers against sputtering damage. They reported comparable OLED external efficiency η_{ext} = 0.1 % with the combination of 100 Å thick Mg-Ag and 400 Å thick ITO to that of the conventional Alq_3-based OLEDs of similar composition with opaque top electrodes. Bulović *et al.* reported an inverted OLED employing novel protective cap layers, 3,4,9,10-perylenetetra-carboxylic dianhydride (PTCDA) and copper phthalocyanine (CuPc), before the sputter deposition of ITO anode with same method like Gu *et al.* [141]. It was demonstrated that these materials can withstand sputter-deposition of ITO with minimal degradation. However, it was also reported that the operating voltage increased and the efficiency decreased. Afterwards, Gu *et al.* reported a full-color stacked OLED consisted of three independently stacked green, blue, and red emission regions [142]. Each electrode consisted of 55 Å CuPc and 570 Å thick ITO layers in each OLED, and the OLEDs had moderate driving voltages and high efficiencies. However, information regarding leakage current and lifetime result were not reported. It is also believed that those protecting organic materials are not suitable for p-i-n type OLED due to their extremely low conductivity compared with doped transport layers. Other groups also tried to deposit ITO on organic layers using various sputtering techniques, but clear comments on device stability such as lifetime were not added [143-145]. The thickness of ITO layers in previous studies was too thin (~500 Å) to supply sufficient charge carriers which are needed to operate a large OLED. In order to get a comparable performance, as if it is operated with an opaque metal electrode, a thick and highly conductive transparent electrode is strongly required, of course, with low damage.

5 Experimental

This section provides information about substrates, materials and instruments used for the experiments. Various OLED sizes and layout types were employed to characterize the OLEDs. Effects of cleaning solutions on the OLED performance are also introduced. OLEDs fabricated were all p-i-n type employing doped transport layers and phosphorescence emitters, and the stacks are described. Equipments and their hardware configurations for the experiments are also introduced, and their purpose and differences in process are also discussed. Also, the instruments used for the characterizations of OLEDs and deposited films are briefly described.

5.1 Substrate layout and preparation

The substrate comprising four quadrant OLEDs is named as 4Q substrate. This 4Q substrate is commonly used at Fraunhofer IPMS for a basic investigation of OLED. In Figure 5.1 the 4Q substrate and a substrate layout comprising eight 4Q substrates and 100 OLEDs are demonstrated. The 4Q substrate size is 35×50 mm², and each quadrant OLED is defined by the opening of passivation. The sizes of the quadrant are 1.245, 4.977, 19.909 and 79.645 mm², and each diode is named as D1, D2, D3, and D4, respectively. In a 100×100 mm² large substrate, hundred OLEDs exist in order to obtain a distribution of OLED characteristics. A substrate layout consisting of eight 4Q substrates and one hundred OLEDs is called Uni-test substrate at Fraunhofer IPMS. Organic layers and cathode are deposited through different shadow masks, and deposition areas on the 4Q substrate are also indicated by different lines in Figure 5.1.

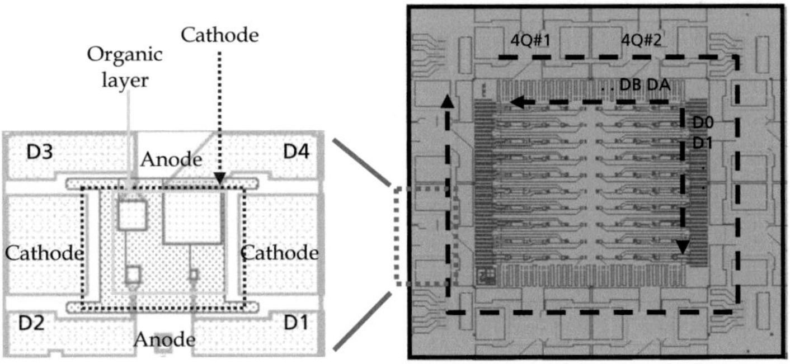

Figure 5.1: Substrate design and layout of 4Q substrate on 200×200 mm² large substrate. This substrate layout is called Uni-test. In the Uni-test substrate, 100 OLEDs having D1 size are located in the middle zone, and eight 4Q substrates are located in the outer zone.

The anode in 4Q substrate is a 90 nm ITO layer for transparency. The 4Q substrate can be, therefore, used for bottom-emission and transparent OLED. The ITO anode has an average surface morphology with 1 nm RMS value and a sheet resistance of 20 Ω/□ with high transparency above 80 % in visible range. For the experiments, either single 4Q substrate or a complete Uni-test substrate can be selected for OLED fabrication depending on the process equipment used. These substrates with the ITO anode are supplied from OPTREX Europe GmbH. For OLED fabrication, 4Q Substrates are cleaned with isopropanol, acetone, and deionised (DI) water ultrasonically. Afterwards they are heated at 120°C for outgassing before they are loaded into the vacuum chamber.

Another substrate type was designed for a top-emission OLED. The substrate is fabricated on 200×200 mm^2 Corning glass without photolithography for film patterning. On a cleaned glass substrate, a 200 nm Ag layer was first deposited through a shadow mask. This Ag layer acts as an opaque anode in OLED. Due to high reflectance (>95 % in visible range) and high conductivity (~ 1.7 μΩ·cm in bulk material), Ag is preferred to other metals for a better coupling efficiency within a cavity structure. The Ag layer was deposited by PK500 magnetron sputter cathode in the Inline system at a DC power of 500 W and a process pressure of 1.33×10^{-3} hPa. The surface morphology and the diffuse reflectance of a 100 nm Ag layer on corning glass substrate were observed by an atomic force microscope Dimension 3100 (Veeco Inc. USA) and UV-Vis spectrophotometer Lambda 900 (PerkinElmer USA), respectively. In Figure 5.2, the surface image and the reflectance of the Ag layer are shown. The Ag layer exhibited irregular grain sizes, and some of them seemed to be sharp. The measured RMS roughness value R_{rms} was 2.218 nm, which is slightly higher than that of the ITO anode in the 4Q substrate. The Ag layer however exhibited superb reflectance in visible range, reaching almost 100 %.

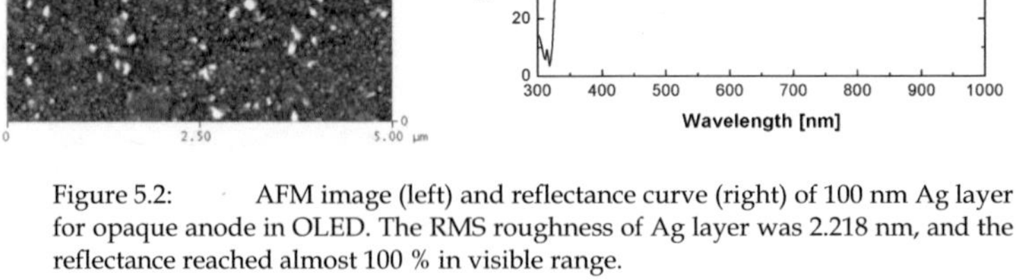

Figure 5.2: AFM image (left) and reflectance curve (right) of 100 nm Ag layer for opaque anode in OLED. The RMS roughness of Ag layer was 2.218 nm, and the reflectance reached almost 100 % in visible range.

In order to define the active area of the OLED, a passivation was formed by means of silk screen printing. Epoxy resin (Peters SD2494 NB-SM) was used for the passivation material, and it was coated on the Ag coated substrate through a stencil silk. Afterwards, the substrates are baked at 130°C for 50 minutes. Since a 200 nm thick Ag film is weak to act as a contact for electrical measurements, a metal paste (Dupont 9145) was additionally coated on the substrates by silk printing. The substrate is then baked again at 130°C for 50 minutes. An illustration of prepared substrate is shown in Figure 5.3. OLED sizes were 36, 16, 16 and 36 mm², named as D1, D2, D3 and D4 OLED, respectively. On a 200×200 mm² substrate, nine substrates are fabricated.

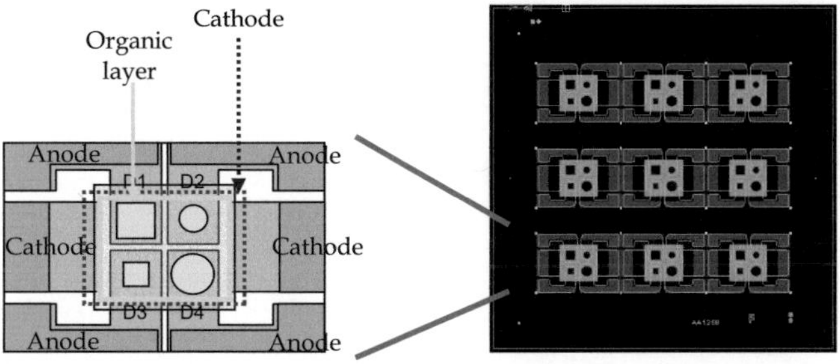

Figure 5.3: Substrate design and layout on 200×200 mm² glass substrate for top-emission OLED.

In case of the substrate with Ag electrode, a careful wet cleaning is necessary in order to minimize a reaction of the Ag surface with chemical solutions.

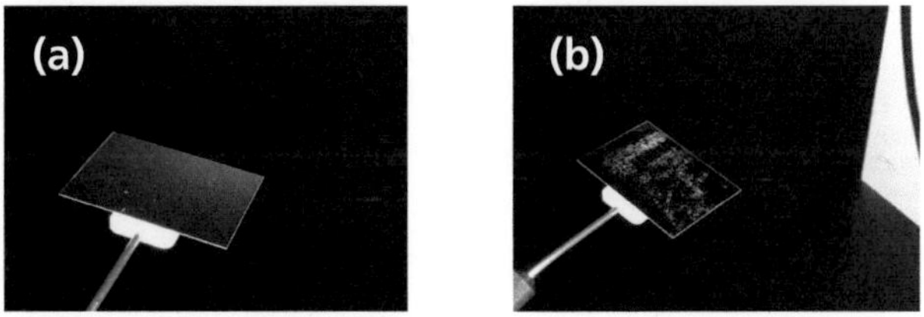

Figure 5.4: Photographs of the Ag film surface under white light illumination. The Ag film was deposited on a metal foil by PK500 in the Inline equipment. Figure (a) is from the Ag film treated by only deionized water, and figure (b) is treated by isopropanol and acetone, namely organic solutions [146].

In Figure 5.4, surface images of Ag films on a metal foil are exhibited. Figure 5.4 (a) is the image after a rinse with only deionized (DI) water, and Figure 5.4 (b) is that after cleaning with organic solutions, isopropanol and acetone. While the Ag film in Figure 5.4 (a) showed a clean surface after rinse, the Ag film cleaned with organic solutions was smeared, showing white spots on the surface. The contaminated surface may influence carrier injection into organic layers, thus only rinsing with DI water in a spin dryer was applied to the substrate cleaning. Before the substrates were loaded in the vacuum equipment, an outgassing procedure was also carried out in a vacuum oven at 120°C for 60 minutes.

5.2 Materials and stacks for OLED fabrications

5.2.1 Organic materials and OLED stacks

Three types of organic stack were implemented to characterize OLEDs after the top electrode deposition by various magnetron sputtering systems. Commonly, doped transport layers were employed in order to efficiently deliver charges to the emission zone with high conductivity and achieve a low driving voltage. As emission layers, two kinds of matrix-guest system were employed for highly efficient phosphorescence orange and green emissions.

OLED comprising electronically doped transport layers are called normally as p-i-n OLED. The basic concept of the doped transport layer was well described by Pfeiffer *et al.* [147]. Experimental study was fulfilled by a co-evaporation of two organic molecules that act as a host and a dopant. For a p-type transport layer, zinc-phthalocyanine (ZnPc) was used as a host material, which is known as a hole transport material, with molecular doping of 2,3,5,6-tetrafluoro-7,7,8,8-tetracyano–quino dimethane (F_4-TCNQ).

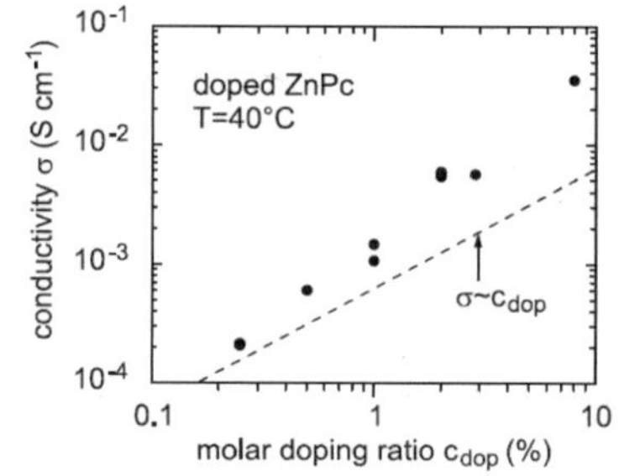

Figure 5.5: Conductivity of ZnPc doped with F_4-TCNQ at 40°C. The conductivity of ZnPc increases superlinearly versus molar doping ratio. The dashed line represents a linear relationship between conductivity and doping ratio. Reprinted from [147].

Figure 5.5 shows the conductivity of doped ZnPc at different molar doping concentrations of F_4-TCNQ. The conductivity increases superlinearly with doping concentration, and it is mainly due to the increase of free carrier density in accordance with the doping concentration [148].

Another advantageous effect is found at the interface between the adjacent electrode and the doped transport layer. In case of the interface between an intrinsic organic layer and electrode, a Schottky contact with a thick space charge region is formed due to the difference of the Fermi level between the organic layer and the electrode. However, if a highly doped organic layer is implemented, an ohmic contact is formed yielding a very thin space charge region. Injection mechanism becomes then tunneling-dominated, thus the injection can be improved. The energy shift was examined by Blochwitz *et al.* implementing ultraviolet photoelectron spectroscopy (UPS) and x-ray photoelectron spectroscopy (XPS) [148].

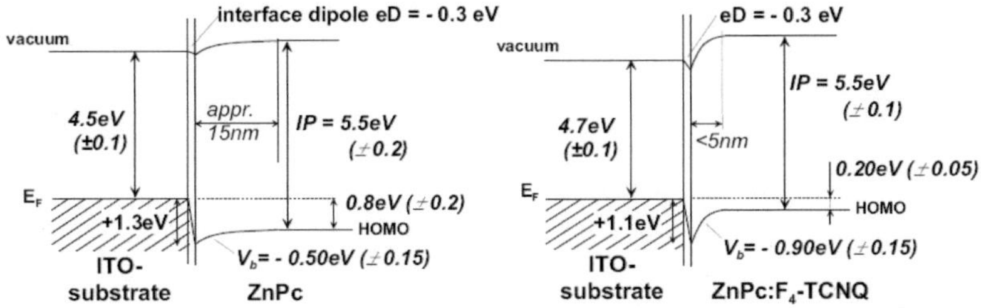

Figure 5.6: Energy diagram of undoped (left) and F_4-TCNQ doped (right) ZnPc on ITO substrate derived from UPS and XPS experiments. Reprinted from Ref. [148].

Figure 5.6 shows the summarized energy diagrams derived from the UPS and XPS experiments. One can see a decrease of the space charge region from 15 nm for undoped ZnPc to below 5 nm for doped ZnPc and a strong bend banding on ITO substrate. In this study, commercial doped transport layers from Novaled AG were used. NHT5 doped with NDP2 was used as a host and a dopant for the hole transport layer, respectively. For the electron transport layer, two host/dopant combinations were used: NET5 doped with NDN1 and NET18 doped with NDN26. Their conductivities were raised by appropriate doping to above 10^{-5} S/cm.

For an emission layer, two host/triplet emitter systems were used for phosphorescence orange and green emissions. The first emission layer consisted of 4,4′-bis [N-(1-naphtyl)-N-phenyl-amino] biphenyl (α-NPD) and Iridium(III)bis(2-methyldibenzo-[f,h]quinoxaline)(acetylacetonate) ($Ir(MDQ)_2(acac)$) as a host and orange-red phosphorescent dye guest, respectively. This emitter system allows an effective energy transfer from the host to the guest dopant because the HOMO/LUMO positions of α-NPD and the triplet energy state of $Ir(MDQ)_2(acac)$ are quite well matched. Hence an effective energy transfer of the singlet and triplet excitons from the host to the phosphorescent dye dopant can be achieved [149]. The second emission layer consisted of TMM004 (Merck KGaA) and fac-tris(2-phenylpyridine) iridium(III) ($Ir(ppy)_3$) as a host and green phosphorescent dye guest, respectively. TMM004 is a commercial electron-transporting material, and it can be also used as the host material for green and red phosphorescent dye guest. Molecular structures of opened materials are demonstrated in Figure 5.7.

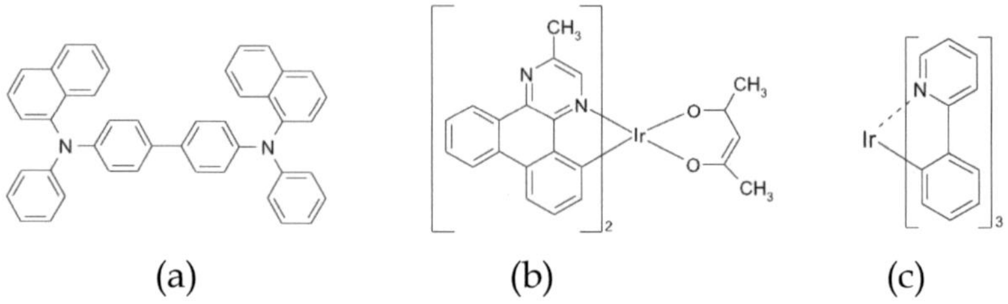

Figure 5.7: Molecular structures of α-NPD (a), $Ir(MDQ)_2(acac)$ (b) and $Ir(ppy)_3$ (c).

Organic layers and their thicknesses for the first organic stack are listed in Table 5.1. This organic stack was deposited on the 4Q substrate for Al cathode deposition on the organic layers. Doped transport layers were used with 10 nm thick α-NPD and NET5 before and after the EML to confine charge carriers for efficiency improvement. As the EML, a phosphorescent orange emitter system was used.

Table 5.1. p-i-n orange OLED stack for Al cathode deposition

Layer	Material	Thickness [nm]
Cathode	Yb or Al	100
ETL	NET5:NDN1	40
HBL	NET5	10
EML	α-NPD:$Ir(MDQ)_2(acac)$	20
EBL	α-NPD	10
HTL	NHT5:NDP2	190
Anode	ITO	90
Substrate	Glass	-

The second organic stack is for phosphorescent green OLED. Organic layers and their thicknesses for the organic stack are listed in Table 5.2. The stack was deposited on the 4Q substrate for Al cathode deposition as well. 10 nm thick α-NPD and TMM004 were used before and after the EML for a reference OLED having 100 nm Yb cathode. On the other hand, NET5 was used as the hole blocking layer in the OLED having an Al cathode. Thus, one can see a difference in efficiency and lifetime resulting from the different blocking layers.

Table 5.2. p-i-n green OLED stack for cathode deposition

Layer	Material	Thickness [nm]
Cathode	Yb or Al	100
ETL	NET5:NDN1	40
HBL	TMM-004(Yb) or NET5(Al)	10
EML	TMM-004:$Ir(ppy)_3$	20
EBL	α-NPD	10
HTL	NHT5:NDP2	190
Anode	ITO	90
Substrate	Glass	-

The third organic stack is for a top-emission OLED on Ag anode. The substrate was already described above, and the organic stack was designed with the software Emissive Thin Film Optics Simulator (ETFOS), which can solve the emission originating from radiative dipoles in a multilayer thin film stack with respect to viewing angle, polarization, layer thicknesses, and refractive indices etc. [150]. The organic stack was calculated for outcoupling in forward direction (0° viewing angle, normal to the substrate) with assuming an isotropic dipole orientation and a dipole location in the middle of the emission layer.

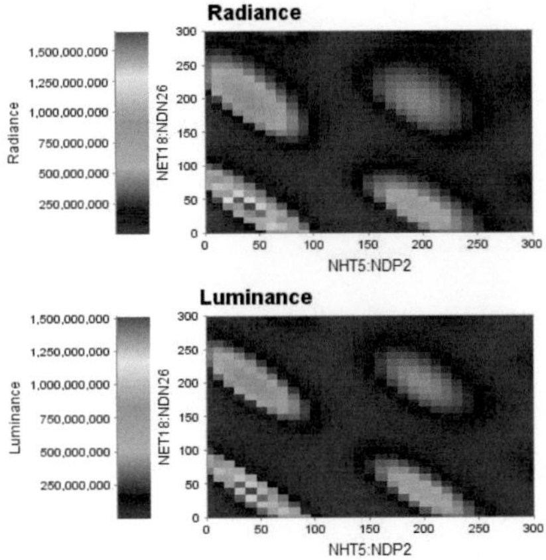

Figure 5.8: Calculated ETFOS results for radiance and luminance on the transport layer thicknesses.

The ETFOS results of radiance and luminance on transport layer thickness are shown in Figure 5.8. Based on the results, the thicknesses of the transport layers were determined to be around 200 nm in order to reduce sputtering damage caused by ITO deposition. The organic layers and thicknesses of the stack are listed in Table 5.3. The thickness of the ETL was in particular varied for a thin and thick layer thickness before ITO deposition on it. Two materials were used for the cathode, and the reason will be discussed in section 6.2.3.

Table 5.3. p-i-n green organic stack of top emission OLED for ITO deposition

Layer	Material	Thickness [nm]
Cathode	ITO	100
Cathode	Al/Ag	3/15
ETL	NET18:NDN26	40, 200
HBL	TMM-004	10
EML	TMM-004:Ir(ppy)$_3$	10
EBL	α-NPD	10
HTL	NHT5:NDP2	198
Anode	Ag	200
Substrate	Glass	-

5.3 Experimental tools

5.3.1 Cluster system

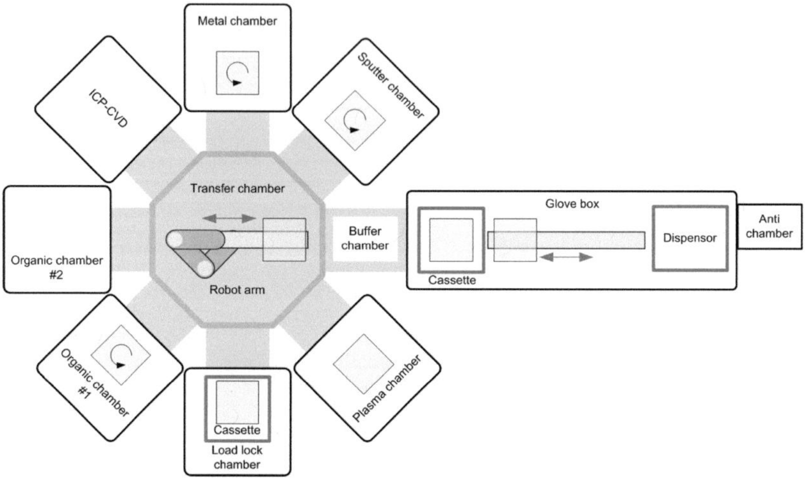

Figure 5.9: Schematic diagram of the cluster type equipment (Sunicel 200, Sunic system. Ltd., Korea) used in this study. It consists of several vacuum process chambers for film depositions and a glove box for OLED encapsulation.

A schematic diagram of the cluster type equipment is shown in Figure 5.9. Sunicel 200 is designed for a 200×200 mm² substrate. Cleaned glass substrates and shadow masks are loaded together in the load lock chamber. After evacuation, each substrate is transported to the process chambers by a vacuum robot arm. Firstly, the substrate is treated by radio-frequency (RF) plasma with Ar and O_2 gas in the plasma chamber. Inductively coupled plasma (ICP) generated by several linear antennas is employed for the treatment due to its high plasma density. This process is for a workfunction modification of the ITO electrode surface as well as for a dry cleaning of organic residues on the surface [151-154]. The substrate is then transported to the organic chambers and the metal chambers for organic and metal depositions, respectively. Each organic chamber contains six thermal evaporation point cells so that a full color or white OLED fabrication is possible in the cluster equipment. The position of the evaporation cells is determined by a computational simulation based on the cosine distribution of evaporant flux from the source. Thus, a good film uniformity of ± 5 % is achieved at a sufficiently large source-to-substrate distance. In the metal chamber, there are two thermal boat systems and three evaporation cells for metal evaporation. Al, Ag, LiF, Ca and Yb are evaporated in the metal chambers. The source-to-substrate distance in the organic chamber and the metal chambers ranges between 450 and 500 mm. The organic and metal chambers are also equipped with auto-alignment systems between the substrate and the shadow mask with an accuracy of ± 3 μm so that a fine patterning of the deposited layer is also possible. This alignment system is in fact essential for OLED display fabrication. The processed substrate is then transported to the glove box via the buffer chamber for an encapsulation with a cover glass and getter in N_2 ambient. Two deposition chambers using plasma are installed in the cluster equipment. The first one is a chemical vapor deposition (CVD) chamber with an ICP type source for a thin film encapsulation. The second one is the sputter chamber in which a linear FTS is installed.

In order to apply the FTS system for large area OLED fabrication, a large FTS is required to cover a large substrate. Therefore, the development of a linear type FTS is required by installing two linear magnetron sputtering cathodes in parallel as reported by Kim *et al.* [7].

The magnetron used for FTS has a specific design compared to a normal magnetron. An important factor to determine the performance of magnetron sputtering is the magnet array and the design. As introduced in Figure 4.8, a normal magnetron consists of a linear magnet array in the middle and a surrounding magnet array with opposite polarity. If FTS is equipped with such a normal magnetron, steep target erosion occurs alike normal sputtering, which results in a low target utilization. However, the target utilization in a FTS system can be improved by realizing a homogeneous magnet field intensity on the target surface. Basically all the magnets used in magnetron are attached on a yoke plate coated with Ni or Cr. An additional yoke plate as well as a compact mount of magnets on a yoke plate can be installed onto the magnets in order to distribute the magnetic field uniformly within the yoke plate. The attachment of such yoke plate on the magnets is avoided in a normal magnetron sputtering due to the loss of magnetic field in front of the target, as explained above. However, the FTS system can use this demerit as a merit, because the magnetic flux is generated between the two yoke plates. Poisson Superfish static magnetic field simulator (Los Alamos National Laboratory, USA) was used in order to calculate a two-dimensional magnetic field distribution. In Figure 5.10, the simulated result of the cross section of the FTS magnetron is shown. The magnetic fields are very homogeneously generated on the targets and in the space between the magnetrons.

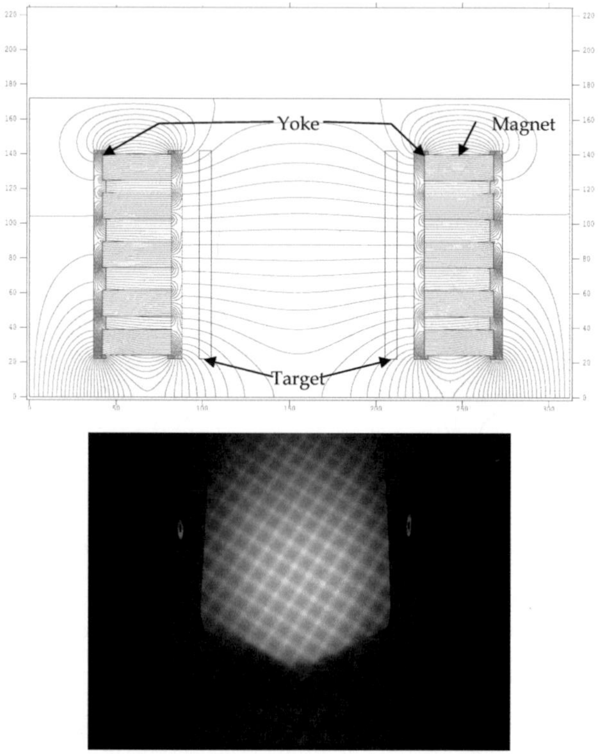

Figure 5.10: Simulated two-dimensional magnetic field distribution of FTS by Poisson Superfish after implementing additional yoke plates on the magnets (Top image). The magnetic fields are successfully formed between the two yoke plates, and the plasma is effectively confined between two targets (Bottom image).

Based on the simulated results of the magnetron, the length and the position of the sputter cathode were determined on the substrate considering the film uniformity deposited on the substrate. A schematic illustration of FTS realized in Sunicel 200 is shown in Figure 5.10 with the electrical diagram of power supply system. RF power supply (Dressler 500 W, Advanced Energy Inc. USA) and DC power supply (Pinnacle 2.5 kW, Advanced Energy Inc. USA) were used for RF super-imposed DC power.

Two target materials were used for the top electrode deposition. One was the pure Al target having 99.999 % purity, and the other was the In_2O_3/SnO_2 (8 wt.% doped) ceramic target having 99.99 % purity. Since they are non-magnetic materials, the generation of the B-field was expected to be identical.

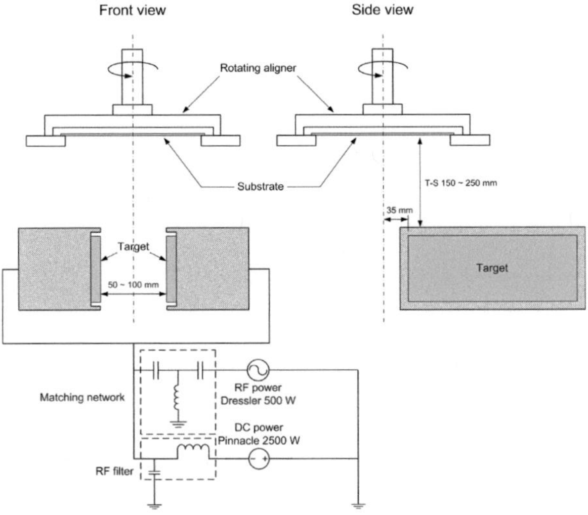

Figure 5.11: A schematic illustration and electrical diagram of the FTS system.

The chamber for the FTS system comprises two linear magnetron cathodes and a rotating aligner where a substrate is mounted. The distance between the targets (T-T) is variable from 50 to 100 mm, and the target to the substrate (T-S) distance is also variable from 150 to 250 mm by electrical motion system. A realized FTS system in a high vacuum chamber is shown in Figure 5.12. A gas delivery system is installed under the process chamber. Two mass flow controllers (MFCs, MKS Co. USA) for Ar and O_2 gases are installed, and their maximum flow rates are 100 and 1 sccm, respectively.

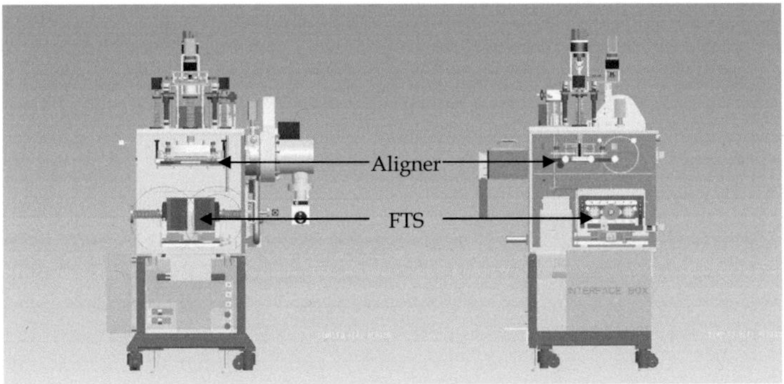

Figure 5.12: Images of the designed FTS system at different viewing angles.

5.3.2 Inline system

The world-wide first vertical Inline equipment (VES400, applied materials GmbH & Co. KG) for OLED fabrication was installed at Fraunhofer IPMS, Dresden [62, 63].

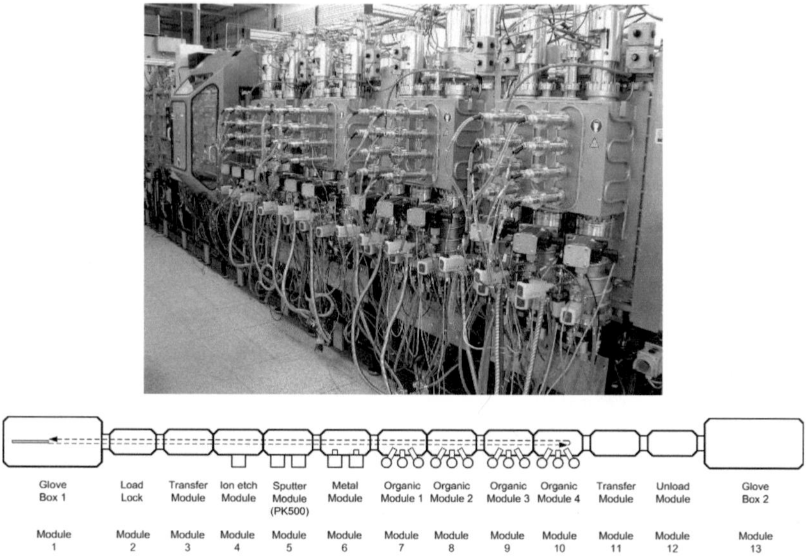

Figure 5.13: Photograph (upper) and schematic diagram (bottom) of the vertical Inline equipment VES400 installed at Fraunhofer IPMS, Dresden.

The VES400 Inline equipment was designed to cover substrate sizes up to 300 × 400 mm². It consists of several vacuum chambers, as demonstrated in Figure 5.13. Since the glass substrate can not move by itself, a substrate carrier is used for transportation. The substrate-loaded carrier is evacuated in the load lock chamber (module 2) and then transported to the ion etch chamber in order to modify and clean the surface of the substrate. A linear ion gun source is installed and is operated under Ar and O_2 ambient at a DC voltage above 700 V. Deposition processes are performed through the whole process chambers (module 5 ~ 10). A unique evaporator design was applied for the vertical Inline equipment. In Figure 5.14, the linear thermal evaporator is shown. It consists of one central crucible and a tube system to guide the vaporized organic molecules. The crucible and the tube are heated up to an adjustable temperature. The heating of the crucible controls the evaporation rate, and that of the tube is to prevent the condensation of the vaporized materials on the inner wall of the tube [108]. A set of maximum three linear type evaporation sources is installed in each process chamber (module 7 ~ 10) to deposit organic materials, which enables doping by means of co-evaporation. One source is used for host material deposition, and another two sources are for dopant material evaporation. After organic layer deposition, the carrier is transferred to metal evaporation chamber (module 6) to deposit a metal top electrode. In the current Inline equipment, Yb and Ag are available for the top electrode. A sputter chamber was installed in VES 400 comprising two linear magnetron sputter cathodes (PK500) as shown in Figure 5.14. Al and Ag target with an area of 87 × 488 mm² were installed for this study. The fabricated OLEDs are encapsulated with getter-attached glasses in a nitrogen ambient glove box, and they are taken out of the glove box for measurement.

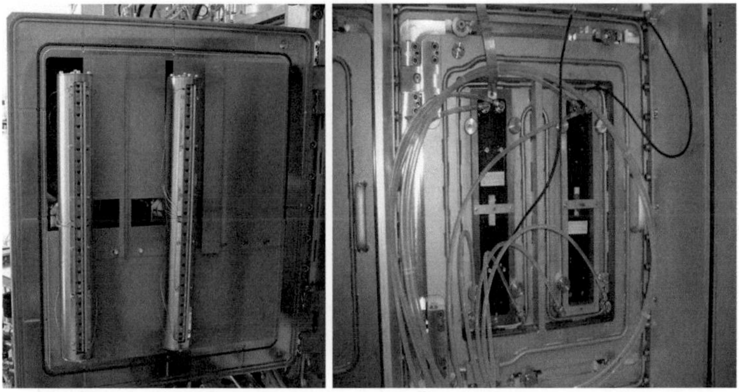

Figure 5.14: Photographs of linear thermal evaporator (left) and magnetron sputter cathode PK500 (right) installed in VES400.

Various types of the substrate carrier can be manufactured depending on the substrate size and dimension. For the experiments of this study, a plate comprising thirteen substrate holders was used in the carrier. The holders are for the substrate size of 35 × 50 mm² and are located within 300 × 400 mm². For the experiments on Al deposition (section 6.1), five substrates were loaded in the holders from the position number one to five as depicted in Figure 5.15.

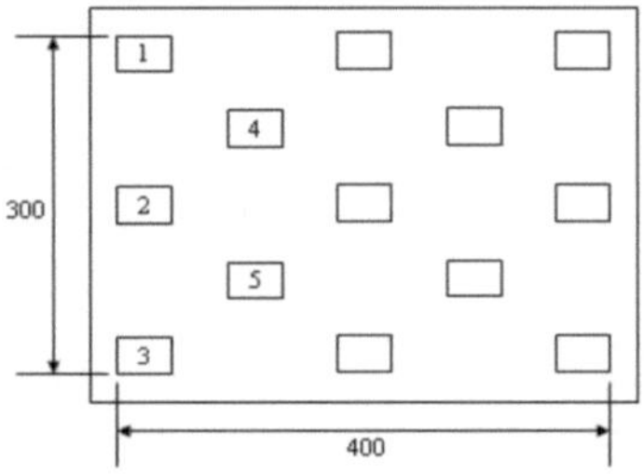

Figure 5.15: Substrate holders and loading positions in the carrier. The carrier moves to left towards the process chambers.

5.3.2.1 Hardware configurations for sputtering

For the different power supply systems, the hardware configurations in the inline equipment were varied to make the process suitable for the cathode depositions. For the Al sputter depositions with DC plasma on the organic layers in Table 5.1, the sputter chamber was equipped with an Al target and a DC power supply.

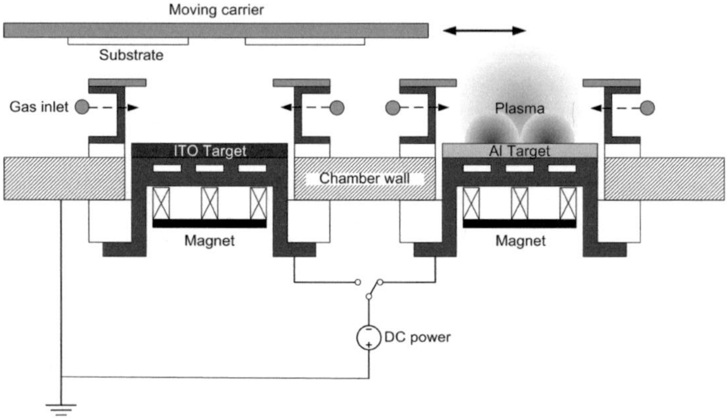

Figure 5.16: The schematic top view of sputter chamber in the vertical inline equipment. The chamber consists of two planar magnetron sputter cathodes. The Al target used for the experiments is installed at the right side.

The schematic of the sputter chamber with only DC power supply is depicted in Figure 5.16. One DC power supply (Pinnacle 6 kW, Advanced Energy Inc. USA) was installed and switched to each sputter cathode so that the ITO target or the Al target is selectively sputtered. Anode shields exist in front of the targets, and the Ar process gas was supplied through the anode shield into the sputtering zone.

In order to realize the RF super-imposed DC sputtering in the inline equipment, one RF power supply (RFPP RF-10S, Advanced Energy Inc. USA) was connected to the sputter cathode with DC power supply. In Figure 5.17, the schematic top view of the sputter chamber is depicted with an electrical diagram. In particular, a RF filter consisting of a capacitance and an inductor was installed for a protection of the DC power supply from high-frequency RF waves. The RF power was coupled capacitively by a matching network, which is necessary to minimize a reflected power by matching impedances of plasma and process chamber.

For DMS deposition, two DC power supplies (GSQ (SA267), Thüringer Leistungselektronik Union GmbH, Germany) were installed for each sputter cathode. A switching unit (UBS-C2, Fraunhofer FEP, Dresden, Germany) was installed between the power supply and the magnetron cathodes to generate unipolar and bipolar square wave pulses. Two pure Al targets were used for Al cathode depositions.

The bipolar pulsed DC power was used for the experiments. The on-time τ_{on} was 9 μs, and the off-time τ_{off} was 1 μs (Duty cycle 90 %) for each magnetron cathode. In every 10 μs the power was switched to each channel (C1 and C2). Therefore the total pulse period was 20 μs. The anode shields at both sides were removed in order to generate the plasma between the two targets. Expected phenomena from the configuration are a wide distribution of sputtered Al atoms with larger angle than in the configuration with the anode shield and a higher deposition rate.

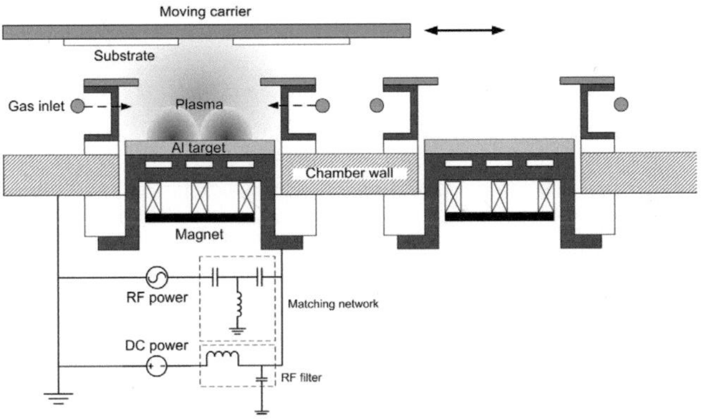

Figure 5.17: Schematic top view of RF super-imposed DC magnetron sputtering system. RF super-imposed DC plasma is realized by a parallel connection of RF power supply and DC power supply. RF filter should be installed between RF and DC power supply in order to protect the DC power supply from high-frequency AC power.

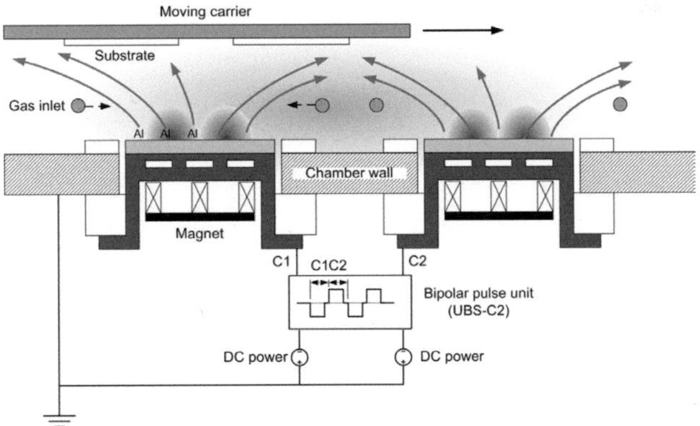

Figure 5.18: Schematic illustration of DMS using bipolar pulsed DC power which is realized in the inline equipment. Blue arrows depict the anticipated wide flux of sputtered Al atoms.

5.4 Characterization methods

5.4.1 Hall effect measurement

The Hall effect was discovered in 1879 by Edwin Herbert Hall [155]. The Hall effect uses the nature of the current consisted of charges like electrons and holes in a conductor. The physical principle of the Hall effect is based on the Lorentz force. Under an electric field, the charges move approximately straight along the electric field. However, when a magnetic field is applied which is perpendicular to their motion, moving charges experience a force, namely the Lorentz force. Thus their path is curved, and the moving charges accumulate on one side of the conductor.

If the conductor is an n-type semiconductor, the carriers are predominately electrons with bulk density n. If a constant current I is assumed to flow along the x-axis from left to right in the presence of a z-directed magnetic field, electrons subject to the Lorentz force initially drift away from the current path towards the negative y-axis, resulting in an excess surface electrical charge on the front side of the sample. This charge results in the Hall voltage, a potential drop across the two sides of the sample. This transverse voltage is the Hall voltage V_H and its magnitude is equal to IB/qnd, where I is the current, B is the magnetic field, d is the sample thickness, and q (1.602×10^{-19} C) is the elementary charge. In usual cases, a sheet density $n_s = nd$ is used instead of the bulk density. The Hall voltage is negative for n-type semiconductors and positive for p-type semiconductors. The sheet resistance R_S of the semiconductor can be determined by the Van der Pauw resistivity measurement technique [156].

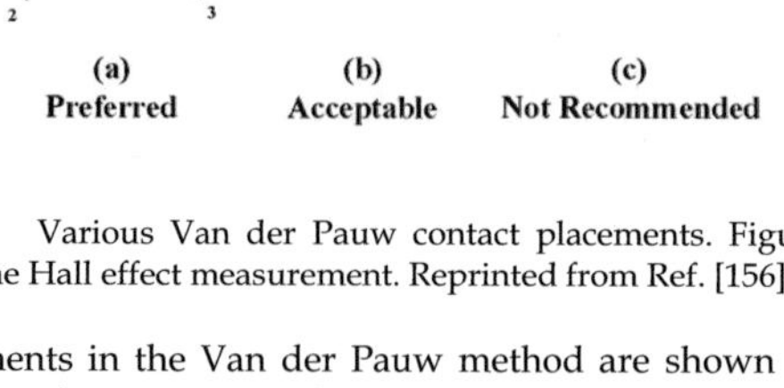

Figure 5.19: Various Van der Pauw contact placements. Figure (b) method was used for the Hall effect measurement. Reprinted from Ref. [156].

Contact placements in the Van der Pauw method are shown in Figure 5.19. For glass substrates, sample preparation for the method (b) is widely used. In order to derive the sheet resistance, the sample is measured twice along a vertical edge, such as $R_{12,34}$, and a corresponding along a horizontal edge, such as $R_{23,41}$. The sheet resistance is then derived from these resistances by the van der Pauw formula

$$e^{-\pi R_{12,34}/R_s} + e^{-\pi R_{23,41}/R_s} = 1 \qquad (5.3.1)$$

Since the sheet resistance is involved with both sheet density and mobility of charges, one can determine the Hall mobility from the equation

$$\mu = \frac{|V_H|}{R_s IB} = \frac{1}{qn_s R_s} \qquad (5.3.2)$$

For the samples of this study, contact placement (b) was used with the sample size of 22×22 mm², and the measurements were carried out under a constant magnetic field of 0.865 T.

5.4.2 Four point probe

Four point probe is a simple instrument which is widely used to measure the sheet resistance of semiconductors and conductors. Basically, it consists of four in-line probes with equal probe spacing. They are connected electrically with voltage and current source meter. The typical scheme for the four point probe measurement is shown in Figure 5.20. By passing a current through the two outer probes and measuring the voltage through the inner probes, one derives the sheet resistance as

$$R_s = F\frac{V}{I} \qquad (5.3.3)$$

where V is the voltage reading from the voltmeter, I is the current carried by the two current-carrying probes, and F is a correction factor.

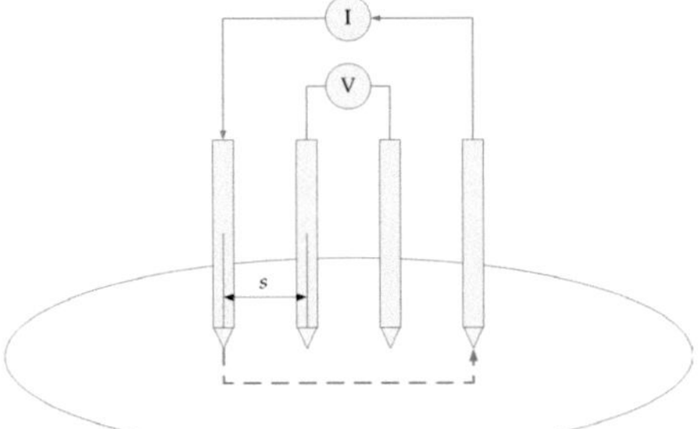

Figure 5.20: Four point probe measurement of a plane specimen.

For probes with equal probe spacing s, the correction factor F is influenced by sample thickness, lateral sample dimensions, and placement of the probes with distances from the sample edges. Normally, a sample size larger than twenty five times the probe spacing s is recommended for a precise resistance measurement. Therefore, 35×50 mm² samples were used for the sheet resistance measurement. The four point probe head with probe spacing 1 mm was manufactured by Jandel, Ltd., and a Jandel HM20 test unit was implemented to read the sheet resistance. The resistivity of the sample was also derived by multiplying the sample or film thickness t to the sheet resistance, based on the relation $\rho = R_s \cdot t$.

5.4.3 Ultraviolet-Visible spectrophotometer

Spectrophotometry is a quantitative measurement of optical properties, such as transmission, reflection, and absorption, of a material as a function of wavelength. Commonly, the spectrophotometry is employed in the UV, visible, and near-infrared regions. The optical properties of thin films are very important in optoelectronic devices so that detailed investigations of each property are strongly required.

Transmittance T is defined as the ratio I/I_0 where I is the intensity of light after the sample and I_0 is the intensity of light before the sample. According to the Beer-Lambert law, absorbance A can be expressed from the transmittance by $A = -\log(T)$ considering zero reflectance and scattering. With the presence of reflection, the absorbance is derived from the relation $A = 1-T-R$, where R is reflectance.

In this study, two kinds of electrode materials are measured by a UV-Vis spectrophotometer Solidspec 3700 (Shimadzu Scientific instruments USA) and Lambda 900 (Perkin-Elmer USA). Sputter-deposited Ag and ITO layers on glass are the samples. The spectrophotometers were available at the Fraunhofer Institute für Photonik Mikrosysteme (IPMS) and Elektronenstrahl und Plasmatechnik (FEP), Dresden for the transmittance and reflectance measurements. The samples were measured at wavelength λ=250 ~ 900 nm with 2 nm step.

5.4.4 Atomic force microscopy (AFM)

Atomic force microscopy (AFM) is a high resolution surface scanning microscopy. It consists of a cantilever with a micro-scale tip, a laser source, and a photo diode. A non-contact mode is used in order to prevent the tip of the cantilever from a contact with a sample surface. AFM (Dimension 3100 scanning probe microscope, Veeco Inc. USA) was used for the investigations of surface morphology. The measurement was performed in the area of 5×5 μm^2. The samples were compared in the peak to valley roughness (R_{pv}), the average (Ra) roughness, and the root mean square (RMS) roughness.

5.4.5 Scanning electron microscopy/Energy dispersive X-ray spectroscopy (SEM/EDS)

The morphology of the sample surface and the degraded area in OLED were observed by scanning electron microscopy (SEM), and the chemical composition of the sample was analyzed by energy dispersive x-ray spectroscopy (EDS).

SEM is an electron microscope which images the sample surface by scanning it with a high energy electron beam in a raster scan pattern. The irradiated sample surface produces signals including secondary electrons, back-scattered electrons, x-rays, light, and transmitted electrons. Since the electron beam is very narrow, SEM has a large penetration depth, which can yield a three-dimensional image. SEM can resolve less than 5 nm so that the technique is very useful to observe a detailed surface structure. When equipped with EDX, an elemental analysis is enabled from the sample simultaneously. The basic mechanism of the EDX is to stimulate an emission of x-rays from a sample by focusing x-rays, electrons, or protons into the sample. The number and energy of the x-ray are measured by an energy dispersive spectrometer, thus the elemental composition of the sample can be measured. The SEM used in this study was JSM-6700-F (JEOL Ltd.), and the EDX was from EDAX Inc.

5.4.6 Ellipsometer

The ellipsometer is an instrument to measure the complex dielectric constant $\varepsilon = \varepsilon_1 + i\varepsilon_2 = (n + ik)^2$ of a given material. ε, ε_1 and ε_2 are the dielectric constants, and n and k are the refractive index and extinction coefficient, respectively. They are all functions of wavelength. The refractive index is used for optics in Fresnel equations and Snell's law, while the dielectric constant is used in Maxwell's equations and electronics. Ellipsometer measure a change in polarization of light upon non-normal reflection on the surface of a sample. Hence a highly reflecting surface is recommended in order to acquire a sufficient amount of the reflected light, and dielectric thin films to be measured are usually deposited on a silicon wafer or a metal thin film.

The incident light is linearly polarized with finite field components E_p and E_s in the directions parallel and perpendicular (senkrecht in german) to the plane of incidence of the light. Upon reflection, the s- and p-components experience a different attenuation and phase shift according to the Fresnel equations. The reflected light therefore is elliptically polarized. The ellipse of polarization of the reflected light is then measured with a second polarizer (analyzer). The complex dielectric constant can be obtained directly from the ellipticity of the reflected light by an inversion of the Fresnel-equations.

In this study, the ellipsometer (DUV-Vase, J. A. Wollam Co., Inc) was implemented in various methods to obtain optical constants of organic materials and electrodes. Especially, the ellipsometry was effectively used to investigate the optical properties of organic materials after a reflective Al film was deposited by magnetron sputtering on them. Due to high reflectance of Al in UV-Vis-IR region, usual optical examinations of the organic material are impossible through a 100 nm thick Al layer. Therefore, a special sample setup was introduced.

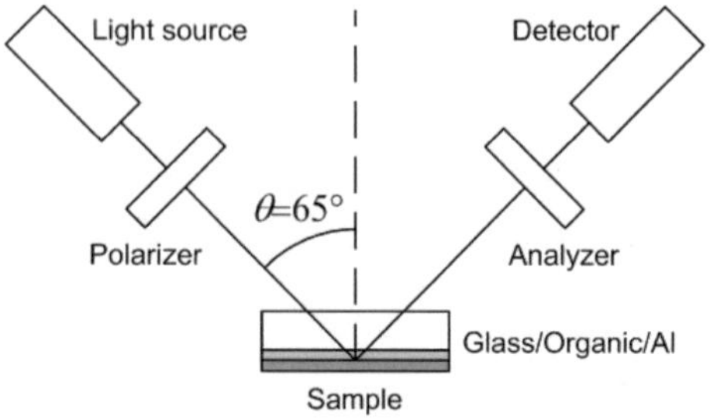

Figure 5.21: Measurement scheme of ellipsometry on the sample having an organic layer and sputter-deposited Al layer on it. The optical constants of the organic layer are obtained from this setup.

As shown in Figure 5.21, a film-deposited glass substrate is set inversely for the measurement. The measurements were performed twice for both sample sides. The incident angle of the polarized light was set to $\theta = 65°$. In the first measurement, the optical constants of the Al film were obtained after eliminating the constant of a thin surface oxide. Afterwards, the film-deposited side of the sample was set upwards like Figure 5.21, and the measurement was performed once more with the setup. In this case, the polarized light was reflected at the Al/organic layer interface and escaped from the sample. The light underwent phase delays by passing through the glass and the organic layer because the refractive indices of glass and most organic materials are higher than air. The light escaped the sample then fell onto the detector so that a complex refractive index was obtained. The optical constants of the organic layer were extracted by means of iterative fitting in the WVASE software. Not only the optical constants, but the thickness of the organic layer can be derived and compared with that measured by profilometer (KLA Tencor).

5.4.7 Laser induced desorption and ionization-time of flight-mass spectroscopy (LDI-TOF-MS)

There are several methods to detect atomic elements in thin films like depth profiling XPS and depth profiling secondary ion mass spectrometry (SIMS). However, these instruments require an additional ion gun in order to sputter the surface of specimen. This is a quite destructive method, and the bombarded surface elements can also be penetrated deeper into the underlying film by collisional energy transfer. Laser induced desorption and ionization-time of flight-mass spectroscopy (LDI-TOF-MS) is a relevant method for the purpose.

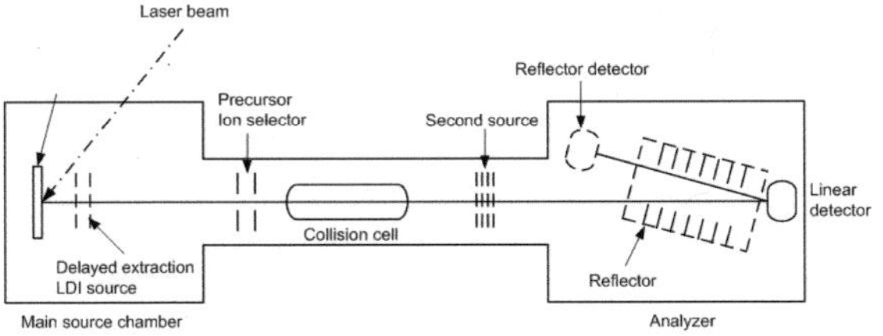

Figure 5.22: Schematic diagram of the LDI-TOF-MS, reprinted from Ref. [157].

In Figure 5.22, the schematic diagram of the LDI-TOF-MS is demonstrated. This method has been used for analyzing polymeric chains and proteins, and it was expended to the OLED application. The LDI-TOF-MS was applied by Scholz *et al.* to predict a degradation mechanism in OLED, e.g. photochemical reactions and influence of organic blocking layer on the luminance decay explaining a reaction of emitter material and the blocking layer [157, 158]. The LDI-TOF-MS was also implemented to observe a shallow doping of metal atoms into organic layer [159]. Thus this method was applied for this study in order to identify a dominant damage source responsible for high leakage current, etc. Usually, the samples with a material of interest are prepared on commercial matrix assisted laser desorption/ionization (MALDI) targets. Since the size of the MALDI target was not adequate for the process equipment, a stainless steel plate with the size of 25×25 mm^2 was used. Since the LDI-TOF-MS can provide in-depth information closer to the steel specimen by laser-induced desorption in deep area, it was expected that the damage source would be discovered. However, the laser intensity should be controlled at a threshold level to avoid an excess of ionizations and corresponding chemical reactions. The samples deposited were placed in the analysis chamber under a vacuum of 10^{-7} mbar during measurement. The wavelength of the laser was 337 nm, and the spectra were taken at positive detection mode at a laser intensity of 90.44 μJ/pulse (75% of the total laser intensity of 129.2 μJ/pulse). Three different positions at one sample were measured, and sixty spectra were gathered from each position to obtain one spectrum.

5.4.8 X-ray diffraction (XRD)

X-ray diffraction (XRD) is an important phenomenon in crystallography to determine a crystal structure of a material. The diffraction is a scattering of X-ray when it strikes regular arrays of atoms, exactly the electrons of the atoms.

The diffraction follows Bragg's law

$$2d \sin\theta = n\lambda \qquad (5.3.4)$$

where d is the spacing between diffracting planes, θ is the incident angle, n is integer and λ is the wavelength of the X-ray.

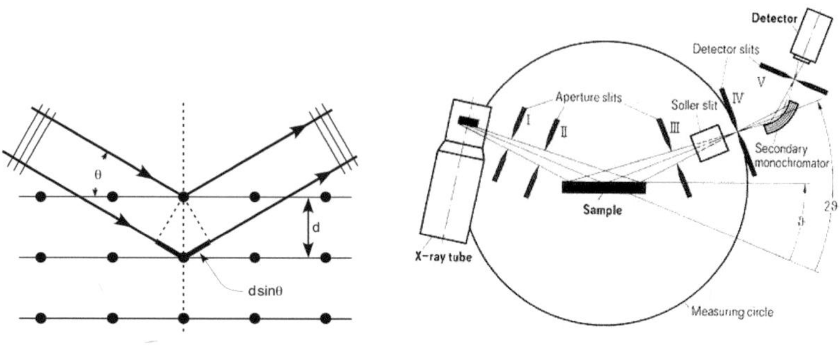

Figure 5.23: Illustration of Bragg diffraction (left) and a basic setup of XRD (right). Incident X-ray is diffracted symmetrically with a spacing d. The difference $2d \sin\theta$ of the beams in path length is an integer multiple of the wavelength λ.

The XRD is used in the crystallographic characterization of single-crystal, powder, and thin film samples. Especially in this study, ITO films on glass substrates were examined by XRD (D5000 diffractometer, Siemens AG). For diffraction, the X-rays of Cu-Kα radiation (λ = 1.54056 Å) were collimated by a Göbel-mirror, which suppresses also Cu-K_β radiation widely, and directed towards the sample. The diffracted beams were collected by scanning the detector arms between $2\theta = 20° \sim 75°$.

5.4.9 Laser interferometer for film stress measurements

The laser interferometer is a thin film stress measurement instrument. It measures the radius of a curvature in a substrate which is caused by a deposition of a highly stressed thin film or a different thermal expansion. Residual film stress is usually classified into two types, namely, a compressive and tensile stress, and the residual stress may result from the contribution of three types of stresses, namely, thermal, intrinsic, and extrinsic stresses [160].

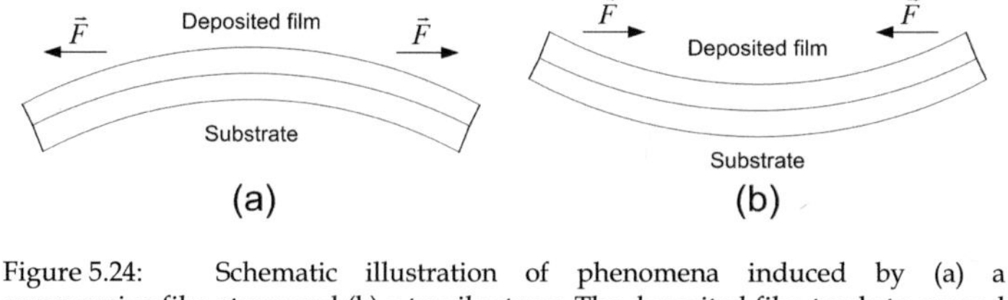

Figure 5.24: Schematic illustration of phenomena induced by (a) a compressive film stress and (b) a tensile stress. The deposited film tends to expand because the film is compressed by the bombardment of energetic particles. The force $\vec{F}$ to release the stress is directed outward for the compressive film stress and vice versa for the tensile film stress because the film tends to shrink.

A schematic illustration of resulting phenomena due to compressive and tensile stress is demonstrated in Figure 5.24. When the film being deposited is bombarded by energetic particles, the film is compressed. Thus, the film tends to expand in order to release the stress, and an expansion force is directed outward the substrate. In opposite, the film tends to shrink because the film is stretched. In this case, the force is directed inward the substrate. The film stress is in particular important if underlying layers are amorphous or exhibit very poor adhesion property with substrate. Therefore, an investigation of the film stress from the top electrode may be important, especially, in a dielectric film deposition, such as ITO.

For the stress measurement, Laser interferometer (FLX-2320, Tencor Instruments) was used, and 6 inch Si wafers having 280 nm thermal oxides were used for film deposition. Before film deposition, the curvature of the substrates was measured first at room temperature. Afterwards, the films were deposited with different process parameters and the curvatures of the film-deposited substrates were measured once more at room temperature in order to distinguish the stress developed during the film deposition.

5.4.10 DMS 101 (Autronic Melcher) for OLED I-V-L characterizations

The *I-V-L* characteristic of OLED was measured by DMS 101 (Autronics Melcher) system, which measures basic characteristics, e.g. electroluminescence spectrum, current density, luminance, and efficiencies, etc.

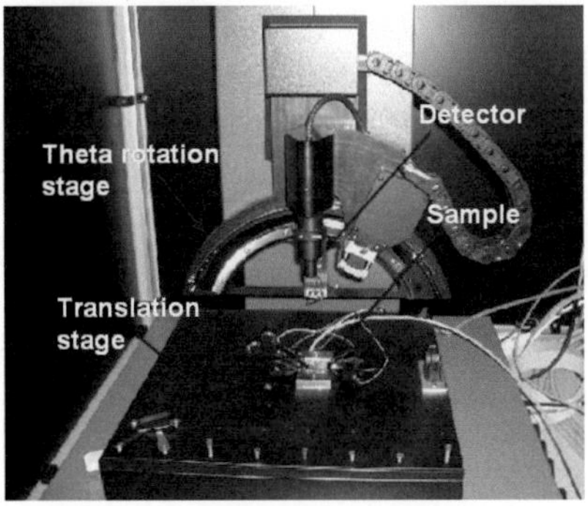

Figure 5.25: Photograph of the DMS 101 (Autronic Melcher GmbH), reprinted from [161].

A photograph of the DMS 101 is shown in Figure 5.25. The emitted light is collected by a photodetector, and the electrical signal is transferred from a source meter (Keithley 2400 sourcemeter). DMS 101 is also equipped with a geniometer which enables an angular dependent measurement. For the OLEDs fabricated in this study, a common voltage regime from 5 to -5 V with a step of 0.25 V was set for the entire measurements. By employing p-i-n OLED, the current density increases dramatically despite a low driving voltage. In order to avoid degradations caused by the high current, the current density was limited at 100 mA/cm^2 compliance for all *I-V-L* measurements.

5.4.11 OLED lifetime measurement

Lifetime of OLED is formally defined as duration for luminance to decay to half of its initial value. The lifetime is evaluated at various operation modes, such as a constant current, voltage, or pulsed voltage modes. Among them, the constant current mode is widely used because OLED is a current-driven electronic device. Also, an elevated temperature can be applied for the lifetime measurements in order to evaluate the stability of the OLED at high temperature or accelerate the degradation of consisting organic materials, enabling a shorter measurement time.

Figure 5.26: Photograph of the lifetime measure instrument OLT (BOTEST systems GmbH, Germany)

The lifetime for the fabricated 4Q OLEDs were measured by the BOTEST system installed at Fraunhofer IPMS. The photograph of the BOTEST system is shown in Figure 5.26. The standard measurement method is to apply specific current densities, e.g. 10, 20, 30 mA/cm^2.

The photodetector measures a relative brightness while the measurement continues until the luminance decreases to under approximate 60 or 70 %. A fitting is then carried out to evaluate the lifetime, and the equation for the fitting is given by

$$\frac{L_t}{L_0} = \exp\left[-\left(\frac{t}{\tau}\right)^{\beta}\right] \qquad (5.3.5)$$

where L_t is a relative brightness at time t, L_0 is a relative initial brightness, τ and β are coefficients for fitting. With the evaluated lifetimes, one can derive the lifetime versus luminance by extrapolating the lifetimes and the initial luminance at each current density.

6 Results and discussions

This section reports the results obtained from various experiments regarding the sputtering deposition of top electrodes on organic layers. Each subsection introduces various process conditions and the corresponding phenomena, which result from magnetron sputtering. The results for the OLEDs fabricated are discussed on the basis of the sputtering mechanisms. In the first section, characteristics of OLEDs having Al cathode deposited on organic layers by DC magnetron sputtering are investigated. The experiments were carried out in the vertical Inline equipment, and they exhibited considerable degradation after the sputtering deposition of the Al cathode. Many experiments contributed to identify the degradation sources. In order to realize a damage-free sputter deposition of the Al top electrode, various power sources and sputter systems were implemented. From the results, it was shown that a combination of a relevant system configuration, a power source, and a process method must be used in order to realize damage-free Al sputter deposition. The next subsection deals with ITO electrode deposition on organic layers as transparent top electrode. For the experiments, ITO films were deposited by the FTS system on thin Ag layer-deposited organic layers or directly on organic layers. Serious degradation was also observed after the ITO deposition. ITO films were deposited with the same process parameters as the Al depositions used FTS. The films were characterized by various methods to investigate the film properties and plasma behaviors. Since the ITO sputtering process is more complicated than Al due to recoiled Ar neutrals and oxygen, the degradation was not completely avoided in spite of a thin Ag buffer layer on organic layers. The results are discussed, and a possible method is suggested at the end of the section, which may enable low-damage ITO sputtering deposition on organic layers.

6.1 Al deposition on OLED using magnetron sputtering

6.1.1 OLED characteristics after Al deposition using DC magnetron sputtering

In this section, the characteristics of OLEDs having sputtered Al cathodes are reported with the results of *I-V-L* and lifetime measurements. Al layers were deposited at various combinations of process parameters based on DC power and process pressure. Since the power and the process pressure determine plasma behavior and most film properties in DC magnetron sputtering, process parameter variation enables anticipation that sputtering damage sources responsible for degradations of OLEDs would be identified. The characteristics of the OLEDs fabricated were compared with those of a reference OLED which has an evaporated electrode.

For the experiments, the VES 400 Inline system (see section 5.3.2) was employed. Five 4Q substrates (see section 5.1) were loaded into the moving carrier for each process after precleaning and outgassing procedures. The p-i-n orange OLED stack in Table 5.1 was used for all the OLEDs, and 100 nm Yb layer was deposited by thermal evaporator on the organic stack for the reference OLED. Al layers were deposited by means of PK 500 sputter cathode at different DC powers with an equal process pressure and at different pressures with an equal DC power. Substrate carrier moving speed, which determines a film thickness deposited during one oscillation in front of the sputter cathode, was held at 0.48 m/min. Only the oscillation count was varied in order to adjust the thickness of the Al layers to 100 nm. Although Yb has a lower workfunction (~2.6 eV) than Al (~4.3 eV), charge carrier injection from both cathode materials is expected to be comparable because the injection mechanism becomes tunneling-dominant ohmic by employing highly doped organic layers.

6.1.1.1 DC Power variation

Several studies tried to achieve a low damage sputtering by supplying very low power [50, 142, 162], but damage seemed to exist nevertheless in form of leakage current and increased driving voltage. In a general sense, it is easily thought that a sputtering with a low power may result in a lower energy of particles produced and, thus, lower damage to OLED. In order to deposit films at different powers, however, different deposition times should be applied for the same thickness. If it is assumed that the degradation level in OLED is determined by an energy transferred by atomic particles to the organic layers during the deposition, a total energy comprising an energy transfer at unit time and total process time should be taken into account. The energy transfer at unit time can be derived from the product of the flux and the kinetic energy of the atomic particles. If the total energy is not considerably different with respect to different powers, the degradation in OLED may show a comparable level.

In Table 6.1, the process parameters, resulting deposition rates, and *I-V* relations from the sputter cathode PK500 upon the process parameters are listed.

Table 6.1.Sets of process parameters for power variations in Al depositions and resulting deposition rates, discharge voltages and currents

Set	Power (W)	Ar flow (sccm)	Pressure (hPa)	Speed (m/min)	Deposition rate (nm/speed·oscillation)	Voltage (V)	Current (A)
1	100	500	5×10^{-3}	0.48	2.48	247.2	0.40
2	300	500	5×10^{-3}	0.48	8.58	283.9	1.06
3	500	500	5×10^{-3}	0.48	14.75	308.8	1.62

Firstly, the energy variation of the particles on the discharge voltages is investigated. As listed in Table 6.1, the discharge voltage increased from 247.2 V to 308.8 V due to an abnormal operation at low power. Incident ion energy induced by the discharge voltage determines the energy of particles ejected from target as well as sputter yield *Y*. Deposition rate, i.e. the number of the particle ejected and condensed on a substrate, is also affected by the number of the incident ions. In order to verify the effect from the voltage difference, the deposition rate and discharge current were indicated with respect to the DC power in Figure 6.1. If the voltage difference significantly affected the sputter yield Y, the deposition rate should have shown a non linear increase with increasing the sputter power. As seen in Figure 6.1, however, the increase of the deposition rate and discharge current showed almost a linear behavior. This means that the influence on the sputter yield by the voltage difference can be neglected and the energy of particles ejected from target is presumed to be comparable.

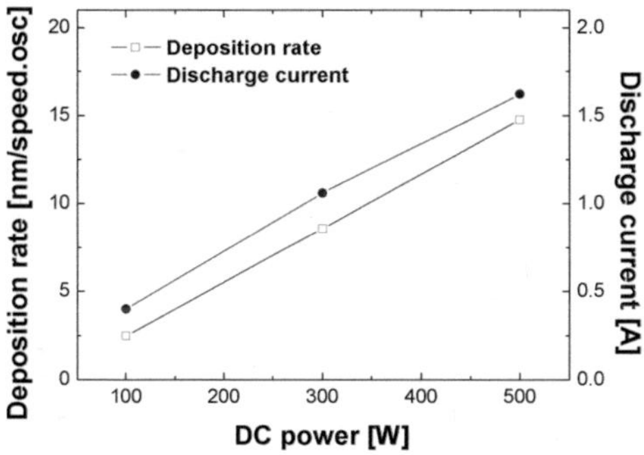

Figure 6.1: The deposition rate and the discharge current with respect to the discharge power in PK500.

Secondly, the flux of the energetic particle is taken into account. The discharge current represents the ion flux at the cathode sheath towards the target surface (see Section 4.2). As mentioned already, the number of the particles ejected and reaching the organic layers at unit time is proportional to the discharge current, if a same collision probability of particles with background gases is given. This condition was provided by the same process pressures. As shown in Figure 6.1, both the deposition rate and discharge current increased nearly proportionally to the DC power. Therefore the instantaneous flux of the particles is supposed to be also proportional to the DC power. For the same 100 nm thick films, inversely shortened deposition times were applied, as the DC power was increased. Since this deposition time represents also the time for the organic layers suffered from the bombardment of the energetic particles, the total energy transfer must consider the kinetic energy, flux, and deposition time.

The total energy delivered by the atomic particles is then derived as

$$E_t = E_k \times \Gamma \times t \tag{6. 1. 1}$$

where E_k is the kinetic energy of each particle, Γ the particles flux, and t the deposition time. The total energy with increasing powers is postulated to be comparable due to the ignorable kinetic energy difference, proportionally increasing flux, and inversely shortening deposition time. As discussed in section 4.2, the sputtered Al atoms are the only atomic particles produced from the target and have high kinetic energy because Ar^+ ions transfer their energy over 95 % to the target atoms and neutralized, generating secondary ions. Therefore, procedures to prove the postulations are required which degradation is caused by the sputtered Al atoms.

Considering the preliminary investigations, the characteristics of the OLEDs having an Al cathode deposited at different powers were observed. One 4Q substrate was selected out of five substrates in each process, and the *I-V-L* characteristics of four 4Q substrates from each process with that of the reference OLED are shown in Figure 6.2.

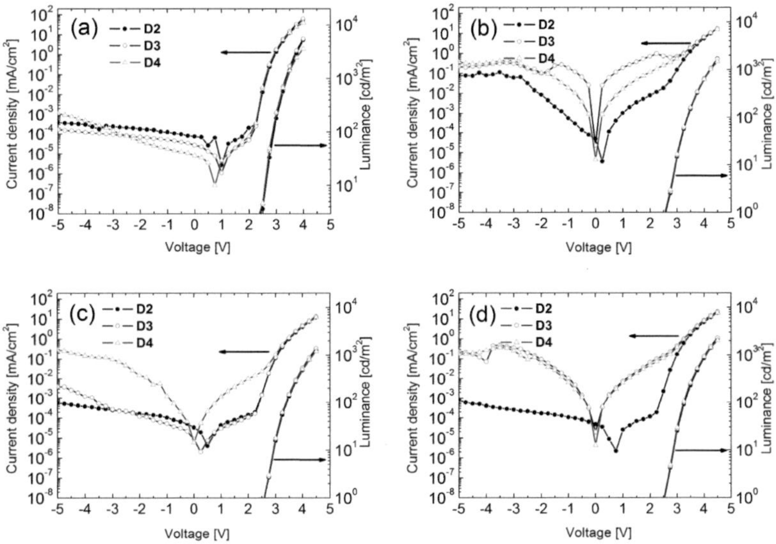

Figure 6.2: *I-V-L* characteristics of the reference OLED (a), OLED with Al layer deposited at 100 W DC power (b), at 300 W DC power (c) and at 500 W DC power (d).

Two interesting features were observed from the *I-V* curves. The first one was that the OLEDs with sputtered Al cathodes exhibit higher current densities at low forward voltage and entire reverse voltage than the reference OLED, and the second one was the decreased forward current densities in every OLED having a sputtered Al cathode. The corresponding luminance of the OLEDs also decreased and the driving voltage shifted accordingly from 3.2 V of the reference OLED to almost 4.3 V of the other OLEDs for 1,000 cd/m². It was also seen that the leakage current densities in the OLEDs ranged randomly from 10^{-4} to 10^{-1} mA/cm² even in one 4Q substrate, while the current densities at the high driving voltage above 3 V were almost identical. Therefore, separate approaches are necessary to identify the damage sources responsible for the two different features.

The higher leakage currents in the OLEDs mean that the OLEDs have additional current paths generated after the sputtering deposition on the organic layers. There might be other possible reasons like a shortcut due to sharp ITO surface or particle contamination, but it is presumably damage by sputtering because identical preparation procedures were carried out for entire substrates. Since the purpose of this section is the stable deposition of a metal top electrode for a large area OLED, only D4 OLEDs out of five 4Q substrates were taken for the comparison of the results. In Figure 6.3 *I-V* and *L-V* curves of the four D4 OLEDs which have been processed at the same position in the substrate carrier are shown.

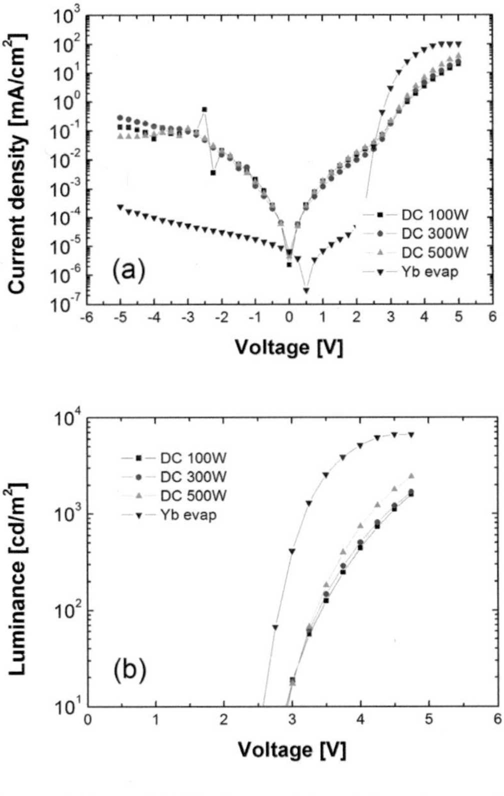

Figure 6.3: Measured *I-V-L* characteristics of the reference OLED and the OLEDs having sputtered Al cathode deposited at various discharge powers. Figure (a) *I-V* curves, and Figure (b) *L-V* curves.

Interestingly, the leakage currents of the OLEDs having sputtered Al cathode showed almost an equal level. This result is quite consistent with the discussion on the total energy transfer. Namely, if the energy difference of the sputtered Al atoms is assumed to be negligible over the small differences in the discharge voltages, the total energy transferred by the impingement of the Al atoms onto the substrate becomes comparable for every discharge power. The exact damage mechanism is, however, still difficult to be established from these results, and will be discussed in the following subsections.

Another remarkable feature was the decrease of forward current density in the OLEDs. As mentioned already, the damage sources responsible for each feature are probably not associated with each other, as shown in Figure 6.2, and only a slightly higher current density was observed from the OLED with Al processed at 500 W DC power. Overall, the forward currents, lying in the space-charge-limited regime, of the OLEDs having sputtered Al cathode decreased and resulted in the corresponding decrease of luminance and, consequently, the increase of driving voltage. Alike the case of leakage current, the responsible damage source can not be explained here and should be discussed in the following section with further experimental results. However, one can anticipate a quite uniform damage source which is apparently distinguished from the damage source of the leakage current.

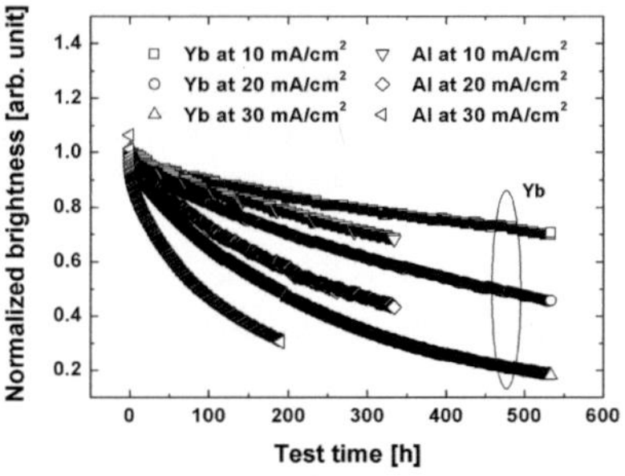

Figure 6.4: Comparison of luminance decay of OLEDs which have the evaporated Yb cathode and the sputtered Al cathode. The OLEDs were measured at the current densities of 10, 20, and 30 mA/cm^2.

Three OLEDs were taken from the reference OLEDs and the OLEDs having sputtered Al cathode, respectively, and used for lifetime measurements. The lifetime was measured at different current densities, 10, 20, and 30 mA/cm^2. It is shown in Figure 6.4 that the luminance decays of the OLEDs having the sputtered Al cathode were much faster than those of the reference OLEDs. This result can be explained with the increased driving voltage as well as the higher leakage current. Driving voltages of the reference OLED obtained from the *I-V* curve in Figure 6.3 (a) were 3.25, 3.45 and 3.6 V at 10, 20 and 30 mA/cm^2, respectively. At the same current densities, the driving voltages of the OLEDs having sputtered Al cathode were 4.2, 4.5 and 5.76 V, respectively. As a result, the OLEDs suffered from a higher electric field such that a faster intrinsic degradation of organic materials might result. The presence of the leakage current paths in the OLED could also lead to such faster degradation of the OLEDs during long time operation, exactly at the places where the leakage currents flow. These currents can induce a continuous local heating, which can deteriorate organic materials much faster than at other places, and, consequently, most of the current flows through these leak points.

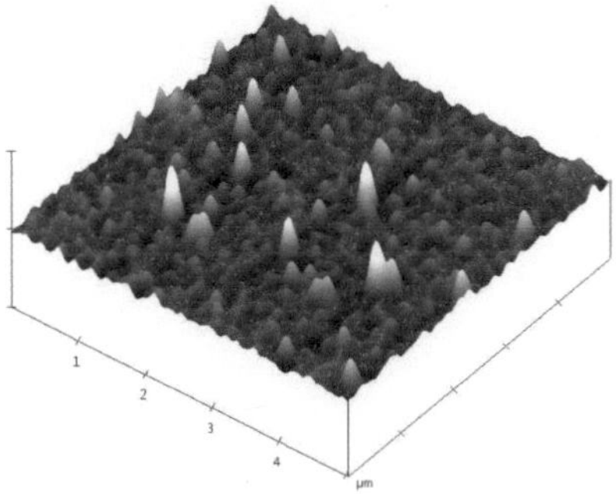

(a)

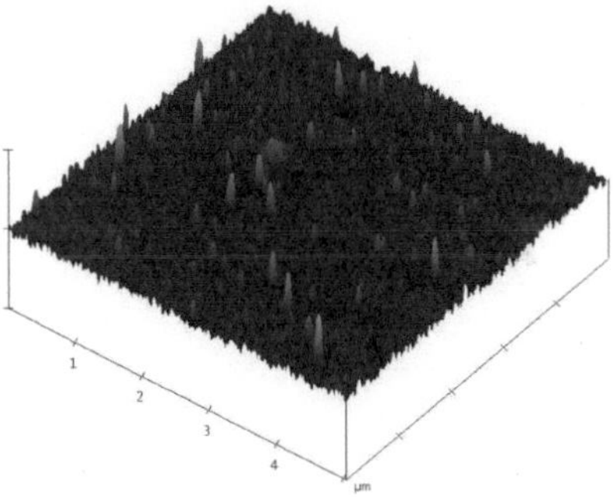

(b)

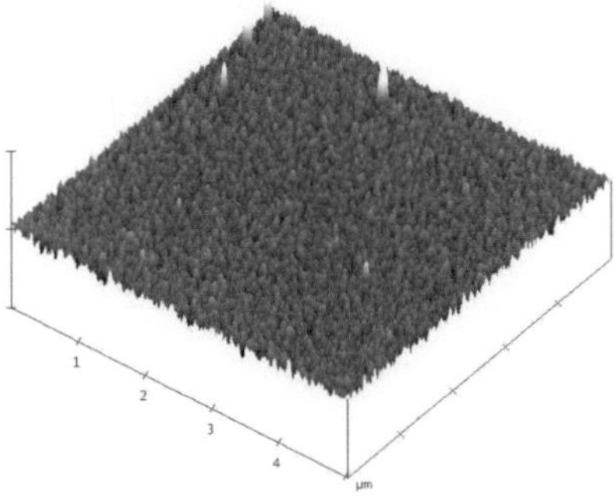

(c)

Figure 6.5: Surface morphologies of Al layers on OLEDs. The AMF images of Al layers deposited at DC power of 100 W (a), 300 W (b), and 500 W (c) are shown. The mean roughness values were R_a = 2.521, 2.736, and 2.728 nm, respectively.

In order to observe the morphology of the Al layers, the surfaces of the fabricated OLEDs were examined by AFM within a 5×5 μm² measure area. The AFM images are shown in Figure 6.5. The average roughness values were 2.521, 2.736, and 2.728 nm for the Al layers deposited at 100, 300, and 500 W DC power, respectively. While the average roughness values showed almost no difference, the formation of hillocks and different grain sizes was observed from the Al layers, showing a gradual change with increasing process power. In Section 4.2, the effect of the deposition rate on the film properties was explained briefly. The thin film growth mechanism by evaporation or sputtering begins with a nucleation of source vapors on a substrate. Low deposition rate results in a lower nucleation density and a corresponding larger grain size followed by further agglomerations. This effect was clearly seen from the AFM images. The grain size of the Al layer was much larger, and many hillocks took place where the Al layer was deposited at a low power and low deposition rate. In contrast, the Al layer deposited at a high power and high deposition rate shows smaller grain size and smooth surface. These results are in accordance with the growth mechanism and can be explained by a stress relaxation which exists intrinsically in a columnar growing layer, because a possible extrinsic stress, e.g. energetic particle bombardments, is supposed to be similar. However, such a morphological difference is believed to play little role in a degradation of OLED, but electrical and optical properties of the Al layer at the metal/organic layer interface may influence the OLED performance, such as injection and reflection of generated light.

6.1.1.2 Pressure variation

In this section, the Ar flow rate was varied for different process pressures maintaining the DC power equally. The process parameters and corresponding discharge voltage, current and deposition rate obtained are listed in Table 6.2. The remarkable point was the decrease of the deposition rate with increasing process pressure, while the discharge voltage and current remained constant.

Table 6.2.Process parameters for the pressure dependent Al depositions

Set	Power (W)	Ar flow (sccm)	Pressure (hPa)	Speed (m/min)	Deposition rate (nm/speed·oscillation)	Voltage (V)	Current (A)
4	500	100	1×10^{-3}	0.48	24.86	346.1	1.44
5	500	300	3×10^{-3}	0.48	18.70	341.6	1.46
6	500	500	5×10^{-3}	0.48	14.75	341.6	1.46

A collision of the sputtered Al atoms with background gases and the resulting back scattering of the Al atoms affect the transport of the Al atoms onto the substrate. Namely, the process pressure influences the deposition rate significantly, and a shortened mean free path is the main mechanism responsible for more frequent scattering at higher pressure.

Helmer *et al.* have examined the pressure effects in planar magnetron sputtering system at a similar process range [118]. In their experiments, an Al target was used for sputter deposition with Ar process gas, and the process pressure varied from 2 mTorr to 30 mTorr, which corresponds to 2.66×10^{-3} and 39.9×10^{-3} hPa, respectively. In particular, a nonlinear decrease of the deposition rate appeared at the pressure between 1.33×10^{-3} and 1.33×10^{-2} hPa, and the deposition rate showed a linear decrease beyond 1.33×10^{-2} hPa. This result explains that the transport of Al atoms changed from a directed transport to a diffusion transport at over 1.33×10^{-2} hPa.

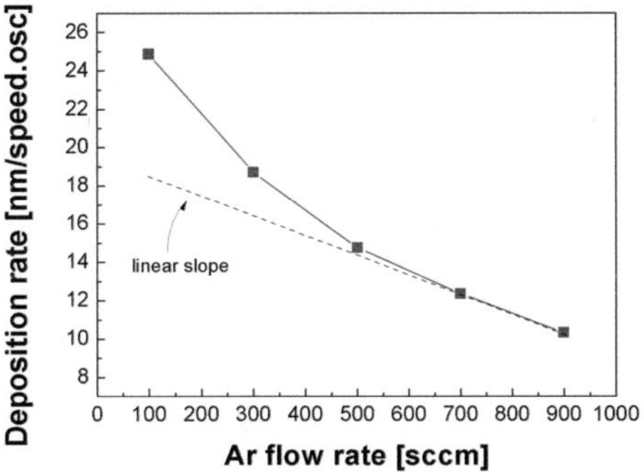

Figure 6.6: Al deposition rates at different Ar flow rates from 100 sccm to 900 sccm. PK 500 sputter cathode in the Inline equipment was used for the experiments.

In the current Inline machine, a pressure over 1.33×10^{-2} hPa could not be reached because the main high-vacuum pump could not evacuate such a high Ar flow rate. Nevertheless, Al depositions have been carried out with Ar flow rates of 700 and 900 sccm to reach a pressure 7×10^{-3} and 9×10^{-3} hPa, respectively. The deposition rates from the Ar flow rate of 100 sccm to 900 sccm are indicated in Figure 6.6. The deposition rate showed a linear decrease after the Ar flow rate of 700 sccm. This linear decrease occurred earlier in the study of Helmer, and this slight difference is supposed to be due to different geometries like chamber volume, vacuum gauge position, or local pressure, etc. These results therefore indicate that the transport of Al atoms may change from a directed transport to a diffusive transport in the Inline machine at the Ar flow rate of over 700 sccm.

Alike the OLED fabrication in Section 6.1.1.1, the thickness of Al cathodes was 100 nm, and only the oscillation count was altered to deposit Al films for the target thickness. In Figure 6.7, the *I-V-L* characteristics of D4 OLEDs taken from the reference OLED and the OLEDs fabricated with Al cathode deposited at various process pressures are shown.

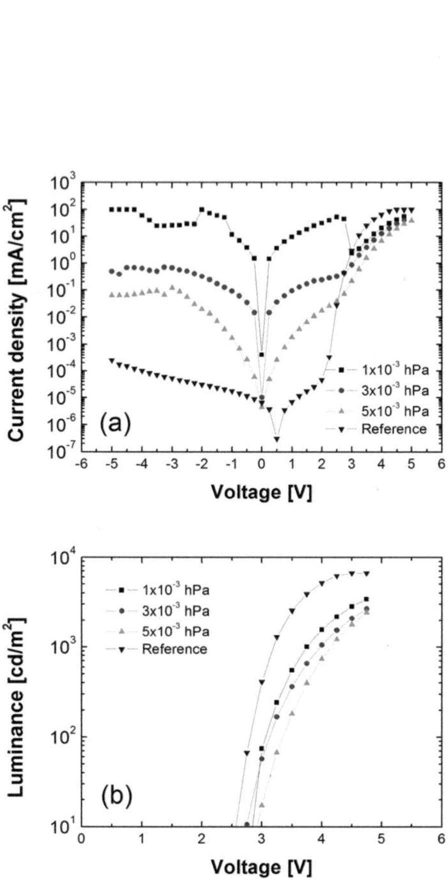

Figure 6.7: Measured *I-V-L* characteristics of the reference OLED and the OLEDs having sputtered Al cathode deposited at various process pressures. Figure (a) shows *I-V* curves, and Figure (b) *L-V* curves.

The remarkable difference among the OLEDs was the variation of the leakage currents at low forward voltage and negative voltage regions. The leakage currents showed a decreasing trend as the process pressure increased. This result provides an important clue that the generation of high leakage current was probably associated with an energy variation of sputtered Al atoms resulting from collisions. The sputtered Al atoms are 'thermalized' through collision and scattering with the background gases, and, consequently, they lose their energy by momentum transfer. Therefore, the pressure variation may enable an identification of a damage mechanism responsible for the high leakage currents. In order to examine the energy variations of particles generated during the magnetron sputtering, the effects on each particle by pressure variation should be investigated.

Among the energy terms of each particle in Eq. (4. 2. 13), only the energy of the sputtered target atom is thought to be affected due to the collision with gases during transport. Since the discharge current is the flow of ions from the plasma sheath to the target cathode, namely ion current, its level is a good indicator for the ion density, the quantity of generated secondary electron and the intensity of photonic radiation near the cathode sheath. Since the discharge current was constant until 5×10^{-3} hPa in the PK500 sputter cathode, the energies and quantities of these particles are supposed to be constant before the collision. In general, the particles except for the ions are less sensitive to the pressure variation because the size and mass of those particles are extremely smaller than atomic particles and the resulting collision cross section is also small for them, which can lead to a much longer mean free path than atomic particles. If a longer process time at higher pressure is applied, the energy transfer by those particles is increased as long as the process time is prolonged. Therefore, an irradiation of those particles is obviously not a cause for the high leakage current.

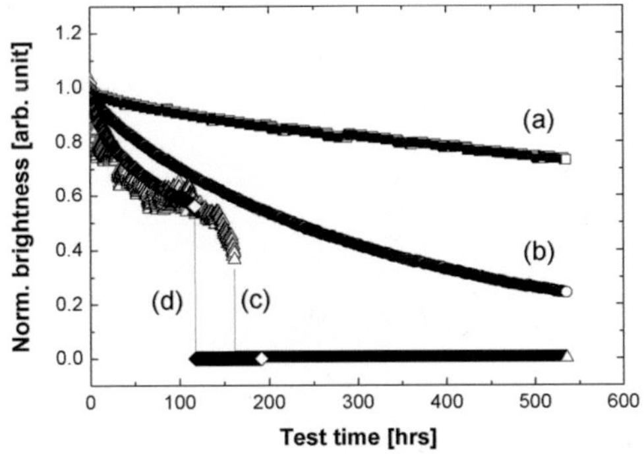

Figure 6.8: Lifetime results of the fabricated OLEDs. Curve (a) and (b) are the normalized brightness of the reference OLED measured at 10 and 30 mA/cm^2, respectively. Curve (c) is the result measured at 10 mA/cm^2 from the OLED with Al cathode deposited at 3×10^{-3} hPa. Curve (d) is the result measured at 30 mA/cm^2 from the OLED with Al cathode deposited at 5×10^{-3} hPa.

Lifetime measurements were also carried out to evaluate the stability of the OLEDs. Figure 6.8 contains four graphs of normalized brightness decay obtained from lifetime measurements. OLEDs, the reference, and the OLED having sputtered Al cathode were supplied with the current density of 10 and 30 mA/cm^2. It should be mentioned first that the OLED with the highest leakage current exhibited extremely short lifetime, only a few minutes, so the result was not included. As expected, the OLED with Al cathode deposited at 3×10^{-3} hPa showed a faster decay, fluctuating until the sample had a sudden death eventually. On the other hand, the OLED with Al cathode deposited at 5×10^{-3} hPa showed a slower decay than the other OLED in curve (d). This result is attributed to the relatively lower leakage current in the OLED. It is therefore manifested that the leakage current and the lifetime have an intimate correlation, and a low leakage current should thus be achieved for a stable OLED with sputtered cathode.

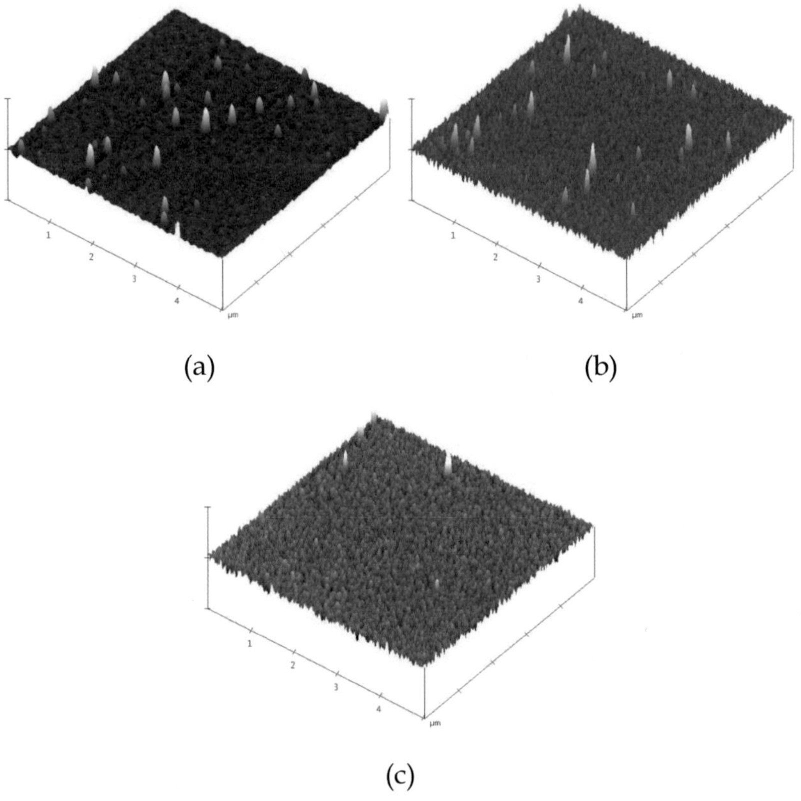

Figure 6.9: Surface morphologies of Al layers processed at Ar flow rate of 100 (a), 300 (b), and 500 sccm (c).

The morphological properties of the Al layers of the fabricated OLEDs were also examined by AFM. As seen in Figure 6.9, the number and the size of hillocks seem to be affected by the process pressure as well. In contrast with the effect of the power variation, a lower process pressure and a corresponding higher deposition rate resulted in a poor morphology. These results can be explained by a bombardment of the energetic Al atoms with higher energy and a stress accumulation in the Al layer. As discussed, the highly energetic atomic bombardment is fulfilled only by the sputtered Al atoms, and the morphology data supports once more that the bombardment by the Al atoms causes a degradation of OLED via energy transfer from the surface of the Al layer.

6.1.1.3 Origin of high leakage currents in OLED

The question which rises from the above experiments is then how the energetic Al atoms can result in the high leakage currents in OLED. Simple structures were made in order to prove and identify the origin. The description of the samples is listed in Table 6.3. The process parameters were taken from Table 6.2 because the Al sputtering depositions by the process pressure variation showed an apparent difference in the leakage current level. The sample A was exclusively made in the Cluster equipment in order to deposit Al by thermal evaporation.

Table 6.3.Sample descriptions for the measurements

Sample	Stack	Process parameter set
A	Glass/NET5(100nm)/evaporated Al(100nm)	×
B/C/D	Glass/NET5(100nm)/Al(100nm)	4 (B), 5 (C), 6 (D)
E/F	Stainless plate/BPhen(20nm)/Al(10nm)	4 (E), 3 (F)

The first work was devoted to examine a variation of electronic band structure of organic material in order to prove a change of the band structure in an organic layer after Al cathode deposition by magnetron sputtering thereon. This intends the verification of the explanation of Liao *et al.* that was described in section 4.5 [5]. Each sample from A to D consists of a single organic layer and Al layer deposited at different process pressures. As an organic interlayer, NET5 (Novaled AG, Dresden) was deposited on the glass because the material underwent the bombardment of Al atoms for the first time. The samples were examined by ellipsometry with a special setup, which was described in Section 5.4.6 in order to obtain optical constants of the materials interested.

Since the optical constants of Al films processed at different parameters were almost identical, the optical properties of the Al film at the organic/Al interface were assumed to be in consistency with that at the air side. Since the optical constant of the glass was already known ($n \cong 1.5$), that of the organic layer was able to be extracted. By means of iterative fits, the thickness as well as the optical constants of the organic layer was finally obtained. The thickness of the organic layer exhibited approximately 5% deviation from the value measured by the profilometer, proving that the sample preparations were very reproducible.

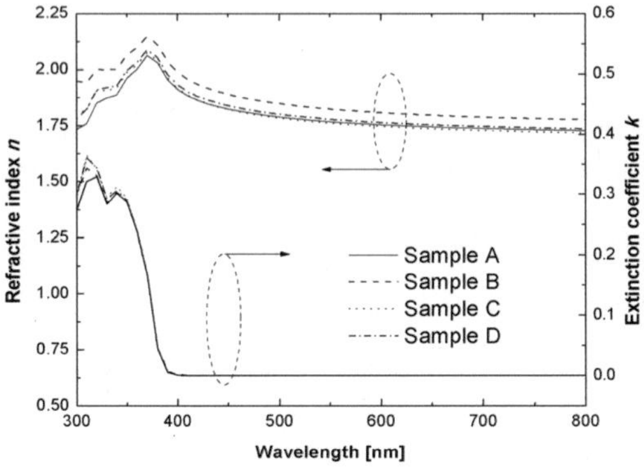

Figure 6.10: Comparison of refractive indices and extinction coefficients of the samples from A to D. The NET5 layers show an identical optical bandgap of 3.26 eV.

In Figure 6.10, the curves of the obtained optical constants of NET5 layers are shown. However, there was no significant difference in the optical properties of the organic layers in spite of the high leakage currents observed from the fabricated OLEDs. In particular, the optical bandgaps of NET5 layers derived from the extinction coefficients were explicitly identical and clearly observable at 3.26 eV (λ = 380 nm); the results were identical in other measured points as well. Semiconductor thin films, which have a forbidden band gap, show usually a long absorption tail by allowing electronic states within the bandgap when the film contains many defects, like lattice mismatch, various polycrystalline phases or impurities [81]. This phenomenon will also be discussed in a later subsection with the result of ITO sputtering deposition. If the organic layer was significantly damaged during sputtering deposition, the layer may contain many defects. Such defects were then able to generate other electronic states within bandgap, allowing absorption at a lower energy than its intrinsic absorption energy. Hence these defects inside of the layer may be detected as a tailing of the absorption curve. This means that the electronic band structure of the entire organic layer was not changed after the Al sputter deposition and the previous explanations from Liao *et al.* are not satisfactory to explain the origin of the leakage currents [5].

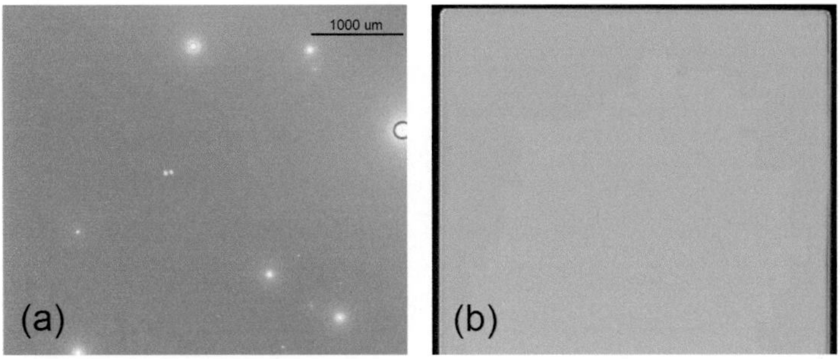

Figure 6.11: Observation of the lighting surface of OLEDs, (a) having sputtered Al cathode and (b) evaporated Al cathode. The driving voltage was 3 V. Many bright spots were found in the OLED having the sputtered Al top electrode while the reference OLED shows a homogenous lighting area.

Besides, the active surfaces of operating OLEDs were also observed by an optical microscope, and the surface images are shown in Figure 6.11. In contrast with the results of Kim *et al.* [8], the OLEDs having sputtered Al cathodes showed many localized bright spots while the reference OLED showed no such spots. These bright spots show that more charge carriers flow through the organic layers due to a high local electric field or parasitic current paths. Such current paths are thought to be responsible for the high leakage currents, and the charge carriers flow not by the conventional hopping conduction over HOMO/ LUMO level, but probably by other highly conductive species.

Such current paths due to conductive species can also incorporate a metallic particle contamination that existed already on the substrate before the layer deposition. Thus, the vicinity and the core of the bright spots were examined by means of SEM/EDS. Figure 6.12 contains two SEM images of the bright spots and a diagram from two spectra of EDS element map observed at the middle of the degradation core and the surface without degradation. It is clearly visible that the middle of the bright spots was seriously destroyed and the total OLED stack at the core was completely molten and agglomerated again. The spectrum of the element map from the degradation core showed only the elements of that the OLED consist. When it is compared with the spectrum from the surface without degradation, no significant difference in elements was detected between the places. Furthermore, these degradations were not observable before *I-V-L* measurement, but generated concurrently with the measurement. This means that conductive and metallic species already existed inside the organic layers, and formed current paths by the applied electric field, letting the current flow directly through them.

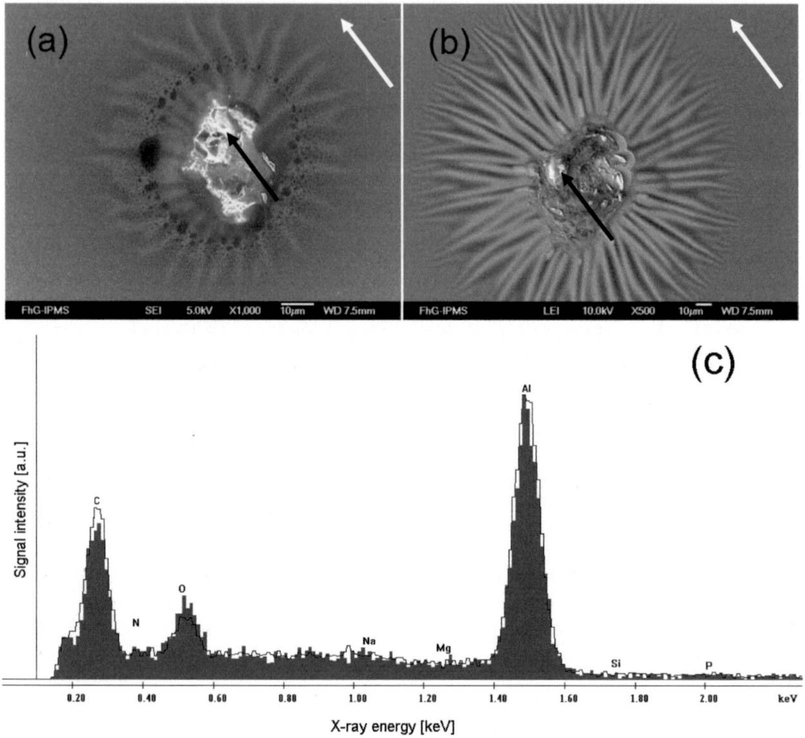

Figure 6.12: SEM images of bright spots after *I-V-L* measurement and diagram of spectra from EDS element map. Figure (a) and (b) show destroyed OLED structures through melting and agglomeration. Diagram of figure (c) shows the spectra of the center of the core (filled) indicated with black arrows and that of the surface without degradation (solid line) indicated with white arrows. Two spectra exhibit almost identical signal intensities from the elements composing OLED.

LDI-TOF-MS was used to identify existing elements and molecules. The samples were prepared on well-cleaned stainless plates. As an organic layer, BPhen was deposited on the stainless substrate. The sample was laid in an analysis chamber with a vacuum of 10^{-7} mbar. The total laser intensity for sample ionization was 129.2 μJ/Pulse, and spectra were taken from positive detection mode at the laser intensity of 75%. Three different positions in one sample were measured, and sixty spectra were gathered from each position to obtain one spectrum.

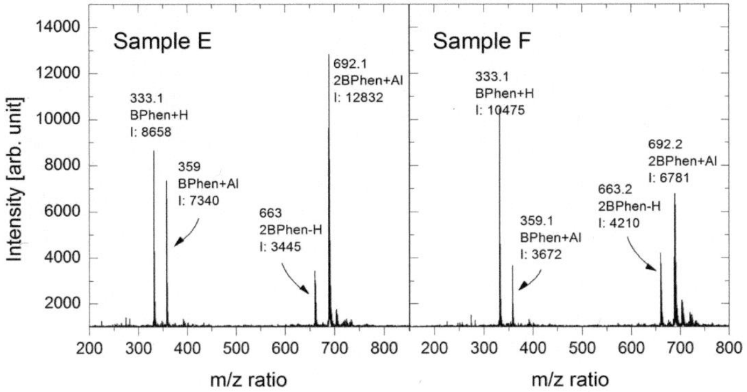

Figure 6.13: LDI-TOF-MS results of sample E and F. The sample E shows much higher intensities of BPhen+Al and 2BPhen+Al complexes than the sample F, representing more Al atoms penetrated to the deeper into the BPhen layer

The organic material BPhen used for the measurements is known to produce a complex with Al or Ag metal atoms by donor-acceptor interaction [159]. If more metal atoms exist in a deep area of the BPhen layer, they will form more complexes with matrix organic molecules. In Figure 6.13, two spectra from the sample E and F are shown. It is obviously seen that the sample E contains more Al adduct ions, $[BPhen+Al]^+$ at m/z = 359 amu and $[2BPhen+Al]^+$ at m/z = 692.1 amu, than the sample F. The ratios of the spectrum intensities from the existing molecules are listed in Table 6.4. This result gives decisive evidence that more metal atoms existed in the sample E than the sample F, supporting the penetration effect of Al atoms due to the bombardment with higher energy.

Table 6.4. Calculated ratios from the spectrum intensities of the sample E and F

Molecules for comparison	Sample E (Ar =100 sccm)	Sample F (Ar =500 sccm)
BPhen/BPhen-Dimer	2.51	2.488
BPhen+Al/2BPhen+Al	0.57	0.54
BPhen/Al+BPhen	1.18	2.85
BPhen-Dimer/2Bphen+Al	0.268	0.62

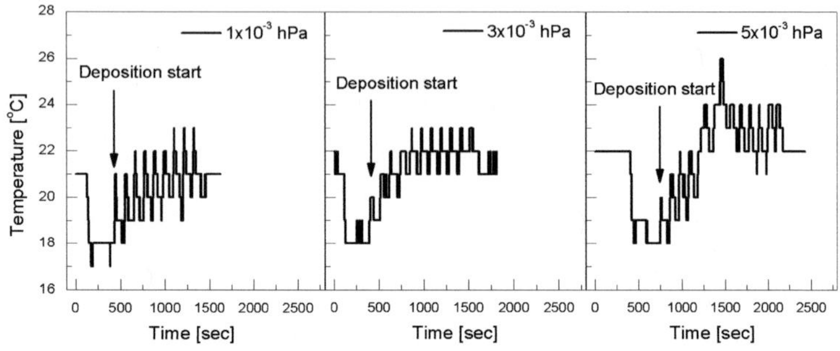

Figure 6.14: Temperature variations during Al deposition at different Ar flow rates. The highest temperature was found as 26°C at the process pressure of 5×10^{-3} hPa which showed the longest deposition time.

A similar phenomenon was reported by another group: even evaporated metal atoms deposited at the substrate temperature of 70°C were able to diffuse into a single crystalline organic diindenoperylene (DIP) film [163]. Accordingly, a high electric field is then induced through the localized paths caused by the metal penetrations, and it allows high local currents through OLED even at a low operating voltage. The substrate temperature during the deposition was measured on a glass substrate by Super mole™ (ECD, Inc.) in order to examine how much thermal energy contributes. In Figure 6.14, the temperature profiles measured during the Al depositions with the process parameters in Table 6.2 are shown. The highest temperature during Al depositions was, however, found to be only 26°C, so the thermal energy may not play a significant role in producing such highly conductive paths through diffusion.

6.1.1.4 Radiation effects

The forward *I-V* characteristics of the OLEDs from Figure 6.7 were plotted again in a linear scale and are shown in Figure 6.15. The reference OLED exhibited a high current density reaching 100 mA/cm^2 at 4 V. On the other hand, the forward current densities of the OLEDs decreased with increasing process pressure. Since the forward current in OLED represents the summation of hole and electron currents, the decrease of the forward current means a decreased conductivity of organic layers or a poor charge carrier injection. It is probably due to a generation of different electronics states or traps at the electrode/organic layer interface or within the organic layers. As discussed in Section 4.2, there still exist other particles in the plasma, which can give rise to fragmentations or reactions of organic molecules. Electrons and photons are those particles that are formally named as plasma radiation, and their energies exceed the binding energy of carbon-carbon covalent bonds or carbon-hydrogen bonds of organic molecules. Proper mechanisms can be introduced to explain the degradation of the forward *I-V* characteristics caused by the plasma radiation: (1) Electrons fall onto a substrate faster than ions when the organic layer is immersed in the plasma. The electrons can be accumulated making the interface of metal/organic layer negatively charged. The accumulation of excess electrons can possibly modify the electronic structure of organic molecules [164]. (2) Photons with energies higher than the band gap of organic material can excite bound electrons to LUMO or higher states, generating electron–hole pairs (EHPs) [165]. When the organic molecules are excited, they may dissipate the excess energy by emitting photons when they return to a lower energy level and eventually to the ground state. The energy can also be emitted irradiatively forming fragmented and reactive species by the opening of a bond structure and the detachment of hydrogen or an organic ligand.

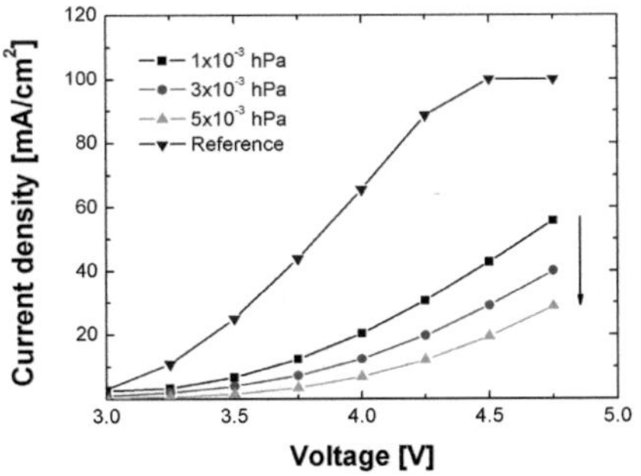

Figure 6.15: Rescaled *I-V* characteristics of OLEDs from Figure 6.7. The forward current decreases as the Ar process pressure increases due to radiative particles from plasma.

As described already in Section 6.1.1.2, the quantity of generated secondary electrons and photons is proportional to the discharge current listed in Table 6.2. Since the discharge current remained almost constant, the flux of electrons and the intensity of photon emission are in principle equal at every discharge power. On the other hand, since the deposition rate has decreased with increasing the process pressure, the surface of organic layer was exposed to the plasma radiation longer accordingly. The generation of the degradations is then related with the radiative energy, which is expressed as a product of radiation flux and exposure time of the organic layer. The radiative energy including E_p and E_e in Eq. (4. 2. 13) becomes then larger at a higher process pressure because the process time to deposit 100 nm thick Al layers for the OLED having Al cathode deposited at 5×10^{-3} hPa was longer than for those deposited at 1×10^{-3} hPa. The resulting effect is clearly seen in Figure 6.15. The forward current decreases from 57 mA/cm^2 down to 30 mA/cm^2 at 4.75 V, as the process time increases. Also, the decrease of the forward current was in good agreement with the increment of the deposition rate. Therefore, the decrease of the forward current can be ascribed to the radiation from the plasma. Additionally, this result assures once more that this radiation-induced degradation should be separated from the origin of the leakage current due to an obviously opposite tendency. The luminance of the OLED was undoubtedly decreased, and this yielded also a shift of driving voltage from 3.2 V of the reference OLED to almost 4 V of the other OLEDs for 1000 cd/m^2.

6.1.2 Al deposition on organic layers using RF superimposed DC power

In this section, Al top electrodes were deposited by RF superimposed DC power. As discussed in section 4.3.2, this technique is quite beneficial to control discharge voltage without losing much deposition rate. If the energy of an Ar^+ ion accelerating toward sputter target, which is strongly dependent on discharge voltage, is decreased, the energy of the ejected particles can be reduced. In a similar experiment by Bender *et al.*, several mixing ratios of RF and DC power were selected [127]. The process parameters and the corresponding discharge parameters are tabulated in Table 6.5. The RF and DC powers were determined by the percentage of the portion of RF power occupying in the total power of 1,200 W. The process pressure was held at 3×10^{-3} hPa with Ar flow rate of 300 sccm. As seen in Table 6.5, the discharge voltage read from DC power supply decreased when the RF portion increased. In an earlier discussion, it has been explained that the decrease of the discharge voltage is contributed to the increase of plasma density resulting from oscillating sheath.

Table 6.5.Process parameters for RF-DC sputter deposition

RF portion (%)	RF power (W)	DC power (W)	Carrier speed (m/min)	Deposition rate (nm/speed·oscillation)	DC voltage (V)	DC current (A)
16.67	200	1,000	0.48	41.67	424.5	2.4
50	600	600	0.48	34.72	268.5	2.3
83.33	1,000	200	0.48	29.76	214.3	0.9

Despite the maximum output power of 1,000 W in the RF power supply, the discharge voltage did not decrease much further than that using only DC power. However, the discharge voltage decreased to almost the half as the RF portion increased from 16.67 % to 83.33 %. Hence, an investigation regarding the variation of the discharge voltage deserves to be performed by fabricating OLED, whether the DC voltage plays a significant role in reducing leakage current of OLED.

In these experiments, likewise, five 4Q substrates were used for OLED fabrication in order to observe the results, such as the distribution of leakage current and forward current, from a large deposition area. The OLED stack was the same as in the previous experiments, and was described in Table 5.1. The encapsulations and measurements were identical as well. The *I-V-L* characteristics from all D4 OLEDs are exhibited in Figure 6.16. Each graph contains *I-V-L* curves from the five OLEDs processed in a same batch. Figure 6.16 (a), (b), and (c) are the results from the Al depositions at the RF portion of 16.67, 50 and 83.33%, respectively.

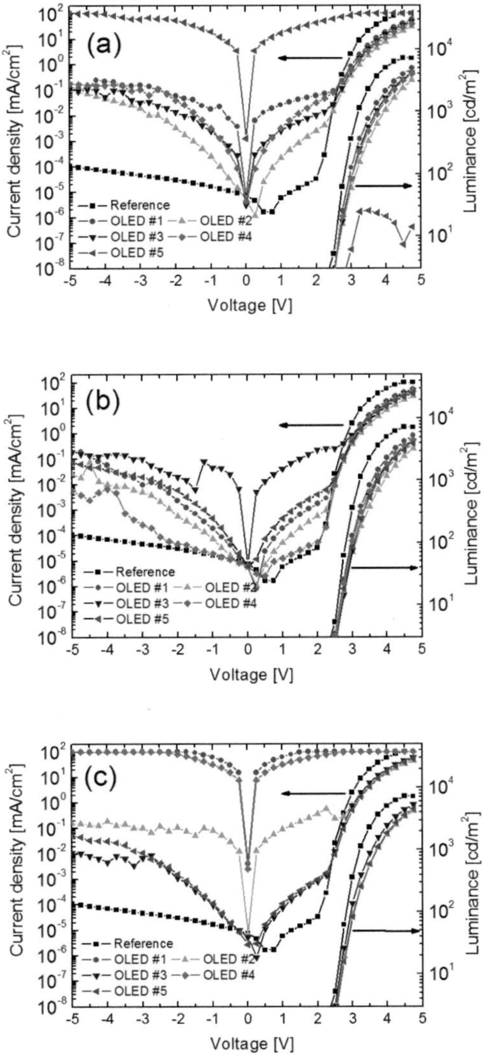

Figure 6.16: *I-V-L* characteristics of the OLEDs. Al top electrodes were deposited at RF portion of 16.67 (a), 50 (b) and 83.33% (c). Each graph contains *I-V-L* curves of five OLEDs

Remarkable points from the *I-V-L* characteristics are again high leakage currents and lower forward currents than the reference OLED. Most leakage currents ranged within 10^{-1} and 10^{-2} mA/cm^2, except for some OLEDs where the leakage currents already reached the compulsory current compliance of 100 mA/cm^2 within ± 5 V. Although the discharge voltage decreased with the help of high RF portion, the leakage currents in the OLEDs were not decreased. It means that the discharge voltage of ~ 200 V was still too high and the resulting energy of sputtered Al atoms could not be reduced significantly. Also, if the slow decrease of the discharge voltage is considered, the discharge voltage may not decrease dramatically. It can also be ascribed to the low Ar flow rate of 300 sccm. Compared with the results of Figure 6.7, a dramatic change of the leakage current level was not found. This means that a shorter mean free path and a resulting thermalization of the sputtered Al atoms by a frequent collision at a higher process pressure was more helpful to prevent organic layers from the Al atom penetration. However, further experiments with higher process pressures have not been performed because the decrease of the forward current was also observed in the OLEDs, and a further decrease of the forward current could be anticipated.

To summarize, RF superimposed DC power sputtering was applied to reduce the energy of Al atoms. Although the discharge voltage was lowered by the addition of RF power, it was still as high as 214 V. The results of the fabricated OLEDs showed high leakage currents as well as decreased forward current. The OLEDs also showed a sudden death problem. Therefore, it is difficult in the Inline tool to realize a damage-free magnetron sputtering with the present hardware configuration.

6.1.3 Al deposition on on organic layer using facing target sputtering (FTS) system

In this section, a FTS system in cluster equipment was evaluated for the Al cathode deposition with various process parameters. In the experiments with the FTS system, 4Q substrates in the Unitest substrate were used for the evaluation and were measured in the same way like the previous experiments. The same p-i-n orange stack was used for the OLED fabrication, and the organic layers were also deposited on the substrate in a cluster machine. For a reference OLED, 100 nm evaporated Al cathode was deposited, and the Al cathodes by FTS also had the same thickness of 100 nm.

In section 6.1.1.4, a radiation-induced degradation in the forward current was discussed. The radiation in normal magnetron sputtering comprised UV photons and secondary electrons. The greatest advantage in using the FTS system is that the secondary electrons and ions do not reach the substrate because the plasma is confined by the strong magnetic field. Thus, only UV photons act as the radiative particle, and the effects of the radiation can be explained with the radiative energy of the UV photons. Influences on OLED efficiencies according to the variation of the forward current were also investigated, because the forward current is a factor which indicates the charge carrier injection and transport, and finally, the carrier balance in the emission zone.

6.1.3.1 Sputter power mixing ratio variation

In this part, results of OLEDs where Al cathodes were deposited by FTS at different sputter power mixing ratio are reported. Five process parameters were adjustable in the current FTS system: RF power, DC power, process pressure, target to substrate distance, target to target distance. The target to substrate (T-S) distance and the target to target (T-T) distance were held at 190 and 100 mm, respectively. The portion of the RF and the DC power in the RF superimposed DC plasma was varied by keeping the total power equally to 600 W. As discussed in section 4.4, this method enables a control of the plasma excitation mode and an energy variation of the generated particles in the bulk plasma.

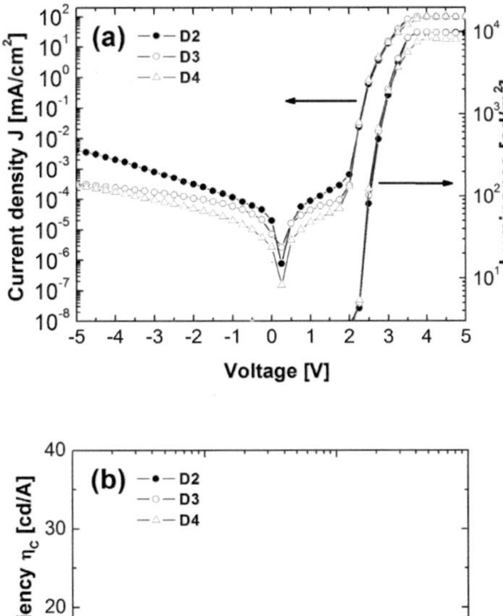

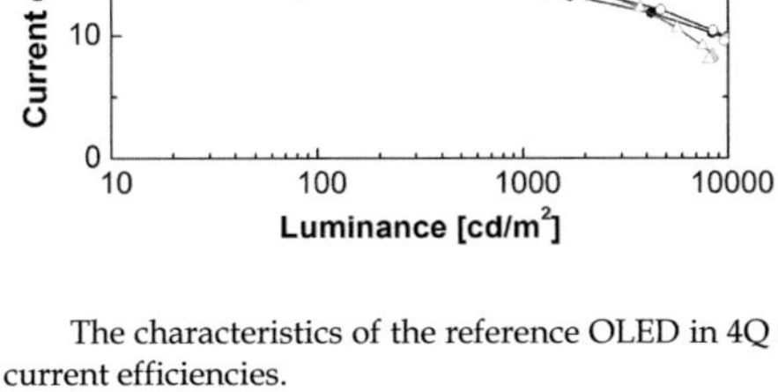

Figure 6.17: The characteristics of the reference OLED in 4Q substrate, (a) I-V-L curves, (b) current efficiencies.

In order to compare the fabricated OLEDs, the characteristics of the reference OLED are addressed first. Figure 6.17 (a) shows *I-V-L* characteristics of OLEDs from one 4Q substrate. OLEDs exhibited quite low current density at reverse bias and low forward bias regions. The compliance of current density was limited at 100 mA/cm² in order to protect OLED from extremely high current, so that the luminance was also limited. The OLEDs exhibit low driving voltages of 2.5 V and 2.82 V in all OLEDs at 100 cd/m² and 1,000 cd/m², respectively. The current efficiency is also shown in Figure 6.17 (b) with respect to the luminance. The reference OLED exhibited 15.38 cd/A and 14.82 cd/A at 100 cd/m² and 1,000 cd/m², respectively. It should be noted once more that this organic stack is indeed not optimized for best outcoupling of the photoluminescence spectrum of the triplet emitter so that a further increase of efficiency is still possible.

Interesting points were the lower forward current density and corresponding luminance of the D4 OLED compared to those of other smaller OLEDs at the same voltage over 3V. This is attributed to a voltage drop across the larger ITO anode and a resulting reduction of current. Since the driving voltage of p-i-n OLED was much lower than OLEDs made of undoped transport layers, the electrodes, especially higher resistive TCOs than metallic cathodes, can not supply a sufficient current at such low voltage.

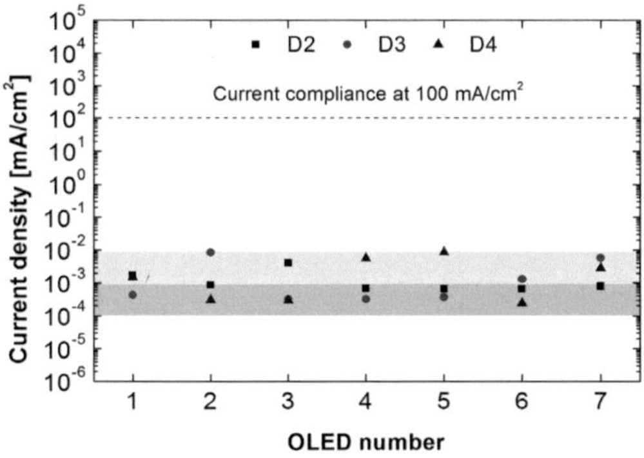

Figure 6.18: Collected leakage current densities of the reference OLED. The OLEDs in the dark gray-filled region between 10^{-3} and 10^{-4} mA/cm² were regarded as degradation-free OLEDs.

In order to observe the level of the leakage currents macroscopically, the leakage current values from the OLEDs measured at –5 V were collected from the reference OLEDs. The values were indicated in Figure 6.18. The reference OLEDs showed very low leakage currents ranging mostly between 10^{-3} and 10^{-4} mA/cm², which lie in a dark grey area. Some of the OLEDs, however, showed a slightly higher leakage current in between 10^{-2} and 10^{-3} mA/cm². Thus an influence of the leakage currents in those areas on OLED stability was investigated through lifetime measurements.

Six reference OLEDs were measured at the current densities of 20 and 30 mA/cm². The corresponding voltage, luminance, and evaluated lifetime are listed in Table 6.6. Furthermore, the percentage of 'sudden death' during the measurement was also included in Table 6.6. The measurements were successful, especially without sudden death problem. It means that the leakage current under 10^{-2} mA/cm² did not affect the long time stability significantly. Lifetime on luminance was obtained by a stretched extrapolation method, plotting the measured lifetimes with respect to the luminance, as shown in Figure 6.19. The extrapolated lifetimes at 500 and 1,000 cd/m² were 47,040 and 12,961 hours. More data points before and after the current densities are probably helpful to evaluate the extrapolated lifetime exactly.

Table 6.6. Results of lifetime measurement tested at different current densities.

Current density (mA/cm²)	Voltage (V)	Initial luminance (cd/m²)	Average lifetime (hrs)	Sudden death (%)
20	3.08	2,599.34	2,439.20	0
30	3.20	3,648.26	1,280.29	0

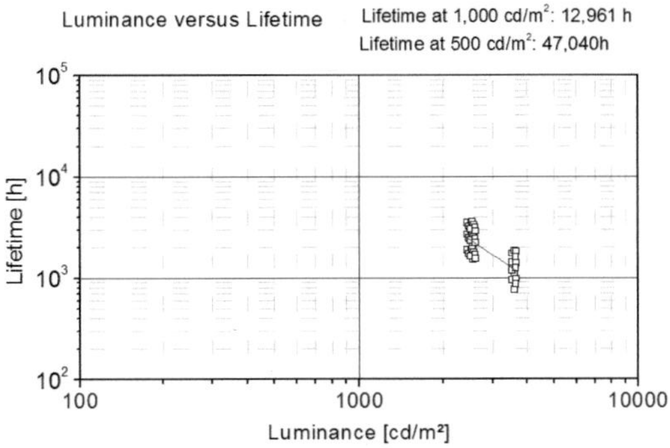

Figure 6.19: Lifetimes of reference OLEDs measured at the current densities of 20 and 30 mA/cm².

For Al depositions, the RF portion was varied from 0 to 83.33 % in four steps. As previously explained, the advantage of power mixture is that the DC cathode voltage can be altered by the mixing ratio. The process parameters and deposition time for 100 nm Al layer are listed in Table 6.7.

Table 6.7. Process parameters for power mixing ratio variation

RF portion (%)	RF power (W)	DC power (W)	Ar flow rate (sccm)	Pressure (mbar)	T-T (mm)	T-S (mm)	Deposition time (sec)
0	0	600	30	3.99E-3	100	190	4,747
16.67	100	500	30	3.99E-3	100	190	5,004
50	300	300	30	3.99E-3	100	190	6,320
83.33	500	100	30	3.99E-3	100	190	9,339

Corresponding results for DC voltage, DC current, deposition rate, and Al resistivity are indicated in Figure 6.20. As the RF portion increased, the DC voltage decreased from 669 V down to 85 V. This phenomenon was discussed in section 4.3.2, and is due to the transition from DC to AC plasma excitation mode and the oscillation of plasma. The transition can occur differently depending on the sputter cathode type, but such a transition seemed to occur at the RF portion around 50 % in this case, because the RF voltage ordinarily alters around ± 200 V in a capacitive discharge. The deposition rate seemed to follow the DC voltage but did not drop considerably like the DC voltage due to the increase of plasma density. Interestingly, the Al film resistivity exhibited a sudden increase over the RF portion of 50 %. If the deposition rate-dependant Al film properties are taken into account, which were discussed in section 6.1.1, the slow increase of the film resistivity at the RF portion from 0 to 50 % is explained consistently with the increase of the defect density by e.g. Hillock formations or grain boundary scattering due to the increase of the process time. The abnormal increase of the resistivity was then presumably caused by an additional effect from ion bombardment in the AC plasma excitation mode, which can induce defects inside the film by an 'atomic peening' effect [166].

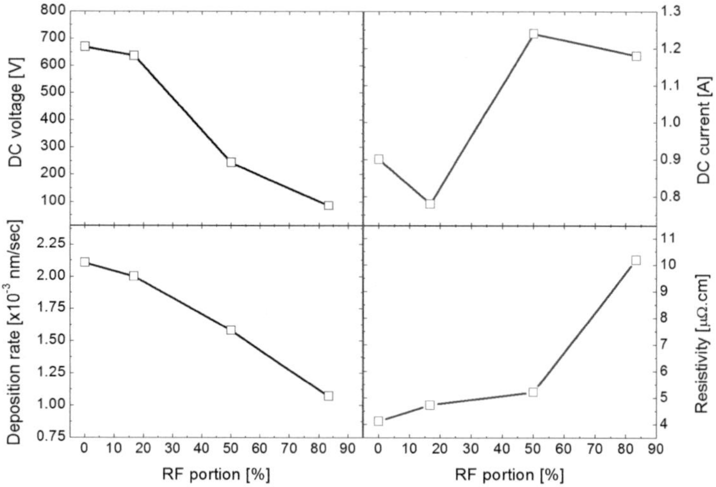

Figure 6.20: Results according to the process parameters in Table 6.7. DC voltage, DC current, Al film resistivity and deposition rate in clockwise. The film resistivity was calculated from the thickness of film and the sheet resistance value measured by four point probe method.

Figure 6.21: Surface morphologies of Al films deposited by FTS (top) and PK500 (bottom) at a same scale. Thickness was 100 nm for both.

The morphology of the Al cathode of the fabricated OLEDs was examined in order to investigate a structural difference depending on sputter systems and process parameters. SEM and AFM were implemented, and the AFM measurement was carried out within 5×5 μm. The surface of the Al layers deposited by FTS in the cluster equipment and PK500 in the Inline equipment was observed by SEM, and the images at the same magnification are shown in Figure 6.21. The first remarkable difference was the uneven grain size in the Al film deposited by FTS. In contrast, the Al film deposited by PK500 showed relatively uniform and smaller grain size with several hillocks thereon. Mechanism on deposition rate-dependent nucleation density, which was described in section 4.2, can be successfully introduced to explain the difference originated from the equipment type. If the flux of the sputtered atoms is high, they collide more often on the surface of the substrate, resulting in a higher nucleation density and small grains like Figure 6.21 (right). On the other hand, larger grains like Figure 6.21 (left) can be formed due to low nucleation density. The rough and inhomogeneous grain structures of the Al layer deposited by FTS are then well understood from the considerably low deposition rate for the same Al thickness.

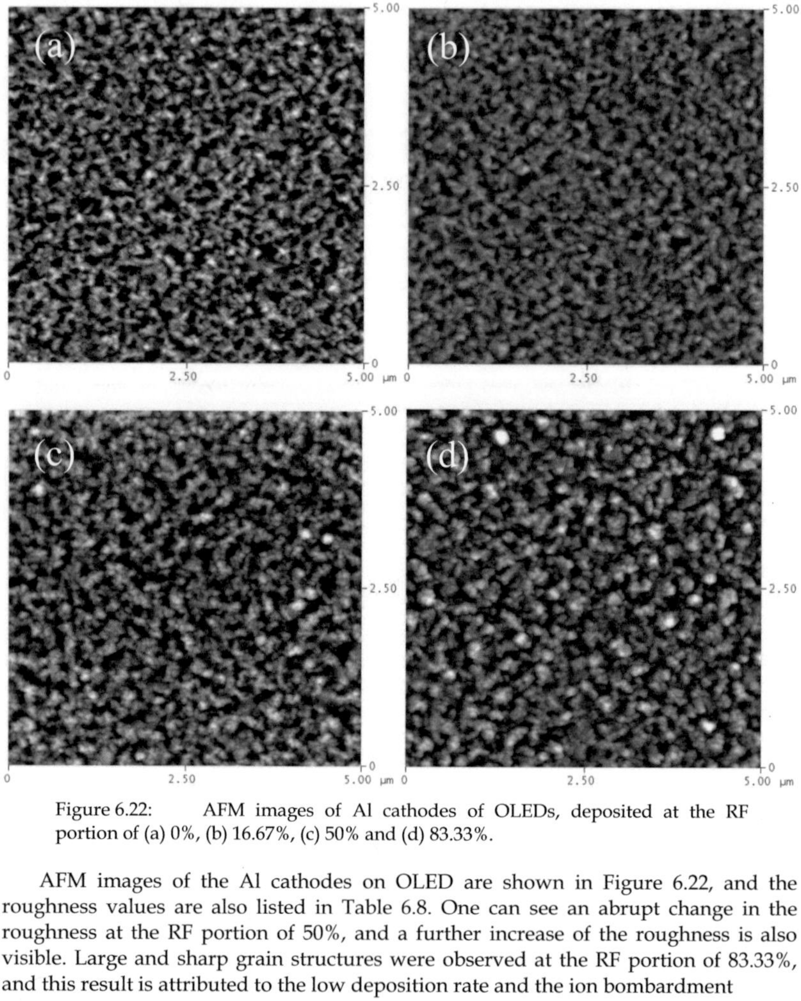

Figure 6.22: AFM images of Al cathodes of OLEDs, deposited at the RF portion of (a) 0%, (b) 16.67%, (c) 50% and (d) 83.33%.

AFM images of the Al cathodes on OLED are shown in Figure 6.22, and the roughness values are also listed in Table 6.8. One can see an abrupt change in the roughness at the RF portion of 50%, and a further increase of the roughness is also visible. Large and sharp grain structures were observed at the RF portion of 83.33%, and this result is attributed to the low deposition rate and the ion bombardment

Table 6.8.Surface roughness of Al cathode on organic layers

RF portion (%)	R_a (nm)	R_{ms} (nm)	R_{pv} (nm)
0	7.752	6.258	61.43
16.67	7.561	6.097	60.10
50	11.35	9.027	93.10
83.33	13.36	10.63	124.89

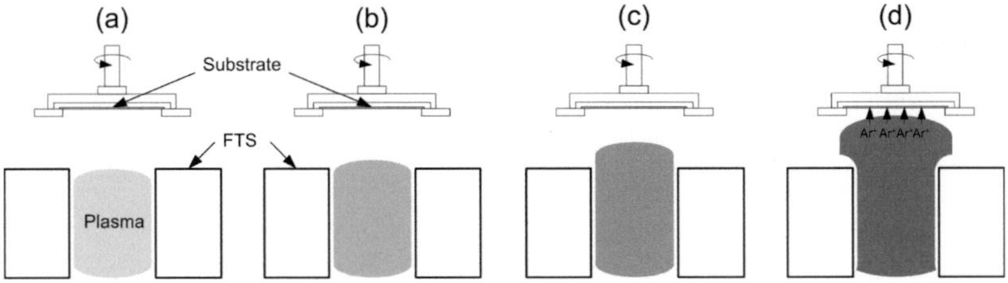

Figure 6.23: Schematic illustration of four plasma excitation modes in FTS with respect to the RF portion of (a) 0%, (b) 16.67%, (c) 50% and (d) 83.33%. Ar+ ion bombardment in the AC plasma excitation mode is also illustrated in figure (d).

Based on the results above, the plasma excitation modes are illustrated in Figure 6.23. The variation of the plasma density was expressed by the color density, and an effective discharge area was also schematized therein. It is emphasized once more that the plasma can spread wider near to the surrounding surface, when electrons start to oscillate with altering AC voltage, and the moving electrons ionize process gases near the surface of substrate. Consequently, the generated ions fall onto the substrate due to the potential difference, and the substrate and the deposited film can therefore suffer from the ion bombardment. One can already anticipate that this bombardment may have enough energy to give a sufficient amount of an impact energy, which can result in metal penetration into the underlying organic layers.

Based on the process parameters from Table 6.7, OLEDs were fabricated and measured for evaluation. *I-V-L* characteristics of four 4Q substrates from each process were selected, and they are shown in Figure 6.24. Figure 6.24 (a) and (b) are showing relatively low leakage currents in all OLEDs, but some of the other OLEDs show high leakage currents. However, they were the result from just one substrate, so the results from the other substrates must all be brought together for the evaluation. Accordingly, all the leakage current values taken at -5 V were collected for an entire overview, and they are indicated in Figure 6.25.

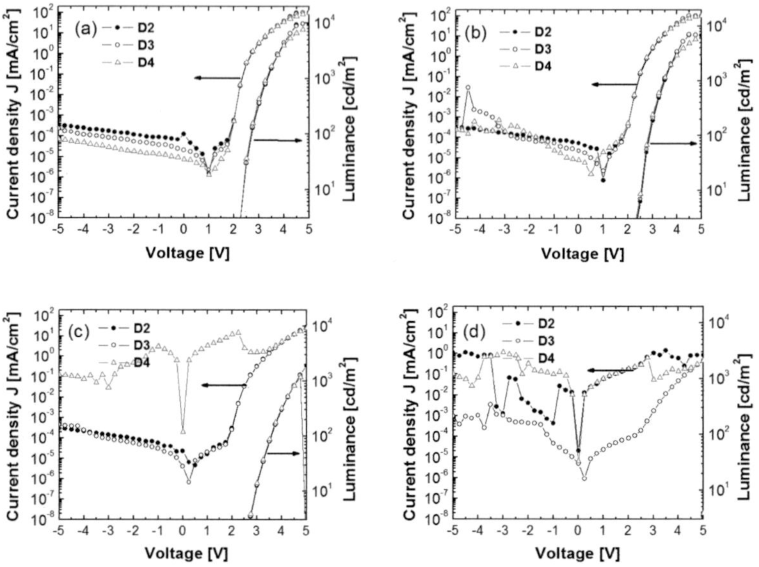

Figure 6.24: I-V-L characteristics of OLEDs that Al cathode was deposited at the RF portion of (a) 0%, (b) 16.67%, (c) 50% and (d) 83.33%.

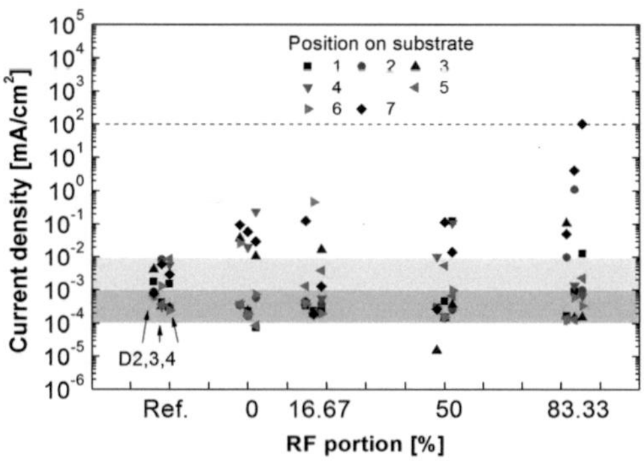

Figure 6.25: Collected leakage current values at -5V from the reference OLED and all the OLEDs having sputtered cathode.

Shown in Figure 6.25 is that some leakage current densities increased steeply at the RF portion of 83.33 %, while the others ranged in the grey area and up to 10 mA/cm². The high leakage currents at 83.33 % can be explained with the ion bombardment effect in AC excitation mode. As seen in Figure 4.13, the impact to the surface of the top organic layer or to the Al film being deposited can be critical because the average energy of ions is as high as several tens of electron volt in the bulk plasma at the high RF portion. The best parameter to reduce the leakage currents on the large substrate was found between the RF portions of 16.67 % and 50%, but those conditions resulted in another degradation induced by radiation.

Unlike the leakage currents, the forward current densities of the OLEDs in one substrate were nearly the same. Figure 6.24 (c) clearly shows that all the OLEDs exhibit almost the same forward current densities at high driving voltage, whereas D4 OLEDs in Figure 6.24 (a) and (b) showed a slightly lower current density. Hence, representative D3 OLEDs from each process were selected for the comparison of the forward current densities.

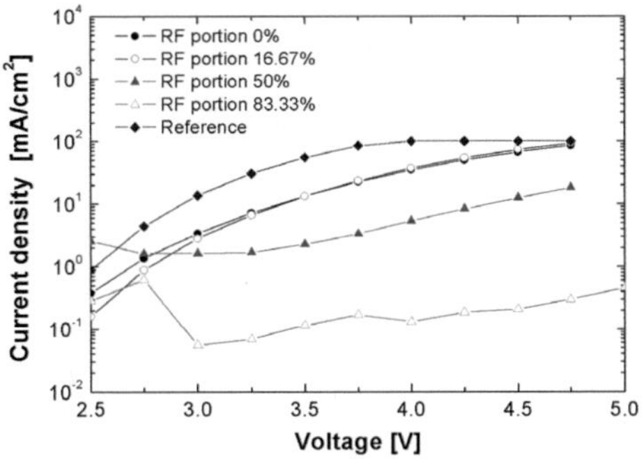

Figure 6.26: Forward I-V characteristics of OLEDs. The forward current density decreases as the RF portion increases.

As the RF portion increased, the forward current density decreased gradually. Especially the forward current density of the OLED fabricated with Al cathode deposited at the highest RF portion was extremely low. The reason for the decrease of the forward current density has comprised several particles in the experiments where only DC power was used. By using FTS, however, the secondary electrons and ions do not reach the substrate in the DC excitation mode because most electrons are confined by the strong magnetic field between the two targets, and also the ions. On the other hand, in the AC excitation mode, the ions can fall onto the substrate due to the oscillating plasma. The rest is then the photonic radiation. The degradation induced by UV photons was also observed from the Al depositions at low RF portions. For that, an optical emission spectroscopy has been widely used for plasma diagnosis, but such a measurement was not performed in the FTS system. As an alternative, an approximation was introduced to calculate the radiation energy with the plasma parameters obtained from the sputter cathode *I-V* relations.

Photonic radiation originates from the energy transition of excited atoms or molecules in plasma, as discussed in section 4.1. The number of excited species can be estimated from the discharge current flowing in the cathode sheath. Since the flow of ions starts near the end of the sheath, it can also reflect the density of ions in the bulk plasma. In DC or low frequency plasma, where the electron oscillation frequency ω is much lower than the plasma frequency, the DC plasma conductivity is approximated as [109]

$$\sigma_{dc} = \frac{e^2 n_0}{m v_m} \tag{6. 1. 2}$$

where $n_0 = n_e = n_i$ is the charge particle density in a cold plasma approximation, m is the mass of charge particle, and v_m is the electron-neutral collision frequency. At a fixed process pressure, v_m stays almost constant, but the charge density can be changed by plasma mode and supplied power. If the DC plasma conductivity is applied extensively to the plasma sheath, the charge particle density n_0 can be derived approximately from a s,heath resistance. The sheath resistance can be calculated from the sputter cathode *I-V* relation of DC power supply. The discharge current is however only readable from the output of DC power supply, so the current from RF power supply was calculated from the DC voltage and the sheath resistance calculated as well. It may be slightly different in the AC plasma excitation mode, since the contribution of electron current is added to the total discharge current. The calculated total currents from the sputter cathode *I-V* characteristics are listed in Table 6.9.

Table 6.9. Total current calculated from RF and DC current.

RF portion (%)	Sheath resistance from DC power supply (Ω)	Current at RF power supply (A)	Total current (A)
0	743.33	0	0.9
16.67	816.67	0.157	0.937
50	195.97	1.24	2.48
83.33	72.03	5.88	7.06

The emission intensity is obtained as

$$I_{emis} = n_e n_s \int \rho_e(E) \upsilon \sigma_s(E) dE = A n_s n_e f(T_e, \rho_s) \tag{6. 1. 3}$$

where n_e is the electron density, n_s is the density of the species of interest (gas atoms, for example), A is a geometry-dependent constant, and $f(T_e, \rho_s)$ is defined as the excitation rate of the species, which is dependent on the electron energy (E) distribution [$\rho_e(E)$] or temperature (kT_e), electron velocity v, and the cross section for excitation $\rho_s(E)$ [167]. The species of interest here are the ions, so the total current listed in Table 6.9 is proportional to the emission intensity. The radiant energy, which the surface of the top organic layer suffers from until a sufficient thick Al is deposited to reflect the UV radiation, was obtained approximately by integrating the intensity with process time, because the thickness for a perfect UV reflection is unknown in the Al layers. The calculated radiant energy is tabulated in Table 6.10.

Table 6.10. Approximated radiant energy calculated from radiant intensity and deposition time.

RF portion (%)	Radiant intensity (arb. Unit)	Deposition time (sec)	Radiant energy (arb. Units)
0	0.9	4,747	4,272.30
16.67 (100 W)	0.937	5,004	4,688.75
50 (300 W)	2.48	6,320	15,673.60
83.33 (500 W)	7.06	9,339	65,933.34

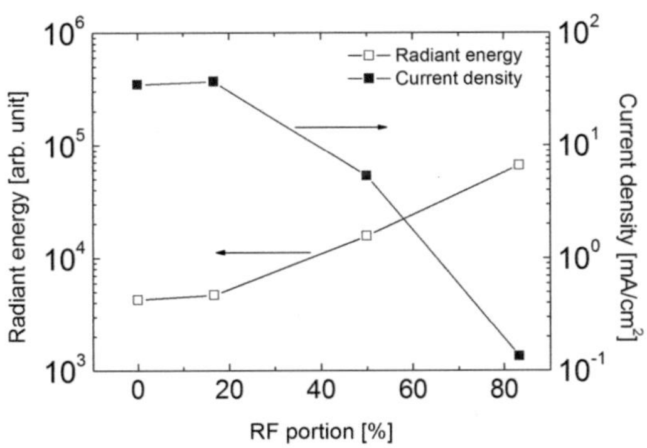

Figure 6.27: Plot of radiant energy and current density at the driving voltage of 4 V.

In Figure 6.27, the values of the radiant energy and the current density taken from the driving voltage of 4 V are indicated. Two graphs are plotted with an equal log scale, and one can see that both plots are reciprocally in a good agreement with each other. One difference was the faster decrease of the current density than the increase of the radiant energy at the RF portion of 83.33%, and it attributes to the effect of ion bombardment from bulk plasma or improved intensity due to contribution of electron current. Nevertheless, this approximation allows the explanation of the VUV radiation effect.

The other effect from the decrease of forward current must then be examined. One can guess a poor transport or a charge injection from the Al cathode, which was caused by a generation of defects inside the organic layers and interface traps primarily at the Al/organic interface. It can then affect exciton generation in the emission layer due to charge imbalance, and the internal efficiency of OLED is significantly influenced eventually. Figure 6.28 shows power efficiency (PE) and current efficiency (CE) of the OLEDs fabricated. The reference OLED exhibited the power and current efficiencies of 16 lm/W and 14.5 cd/A at 1,000 cd/m^2, respectively. However, the OLEDs having the sputtered Al cathode showed gradually decreasing efficiencies when increasing the RF portion. This result reflects the imbalance of charges and the corresponding lower recombination rate in the emission layer very well. The efficiencies with respect to the luminance are also listed in Table 6.11. In particular, the effect of the radiation seems to be extreme in the complete AC plasma mode as seen in the current density and the efficiencies of Figure 6.26 and Figure 6.28.

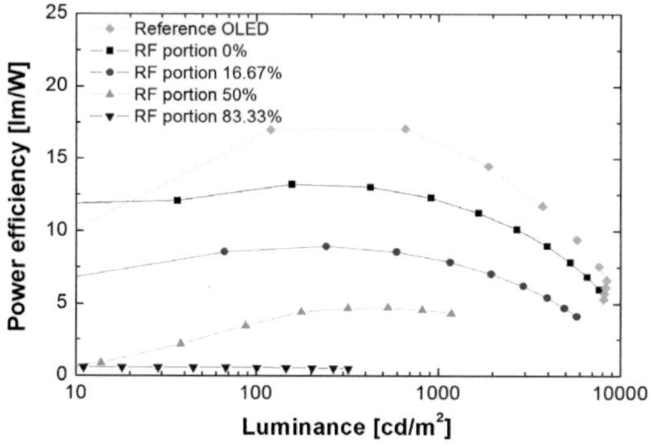

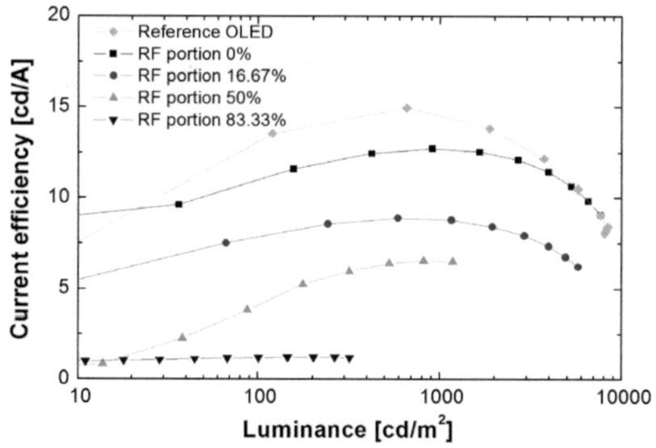

Figure 6.28: Power and current efficiencies of OLEDs. The efficiencies decrease subsequently in accordance with the increase of the RF portion.

Table 6.11. Power and current efficiencies of OLEDs with respect to the luminance of 100 and 1,000 cd/m².

RF portion (%)	PE at 100 cd/m²	PE at 1,000 cd/m²	CE at 100 cd/m²	CE at 1,000 cd/m²
Reference	16.57	16.15	13.18	14.49
0	12.83	12.19	11.00	12.68
16.67 (100 W)	8.68	8.04	7.83	8.76
50 (300 W)	3.68	4.37	4.13	6.49
83.33 (500 W)	0.57	-	1.19	-

Six 4Q substrates from each process were taken for lifetime measurements. Three 4Q substrates were used for each current density of 20 and 30 mA/cm². The evaluated lifetimes and sudden death percentages during the lifetime measurement are indicated in Table 6.12. The OLED processed at the RF portion of 16.67% yielded the longest lifetime among the OLEDs having sputtered Al cathode, but a low sudden death rate at the same current density was found at the RF portion of 50%. In fact, it is difficult to distinguish a best performance in the DC plasma excitation mode, but the highest sudden death rate was found at the RF portion of 83.33%, and it attributes to the high leakage current and high driving voltage, which resulted from the ion bombardment and the high UV radiation, respectively.

Table 6.12. Measured lifetimes of OLEDs

RF portion (%)	Voltage (V)		Average lifetime (hour)		Sudden death (%)	
	at 20 mA/cm²	at 30 mA/cm²	at 20 mA/cm²	at 30 mA/cm²	at 20 mA/cm²	at 30 mA/cm²
0	3.7	3.9	1,309	483	25	0
16.67	3.69	3.86	1,778	1,039	0	66.67
50	4.38	4.6	1,375	516	0	16.67
83.33	7.35	7.68	4	203	100	83.33
Reference	3.1	3.2	3,583	1,758	0	0

6.1.3.2 Sputter pressure variation

In this part, the process pressure was varied from 1 to 5×10^{-3} mbar at two different process powers. One was the combination of RF 300 W and DC 300 W (RF portion of 50%), and the other was only DC 600 W (RF portion of 0%). The process pressure influences the collision frequency between particles due to the change of mean free path, and this effect was expected to alter the energy of the sputtered Al atoms. The process parameters are indicated in Table 6.13.

Table 6.13. Process parameters for pressure variation

set	RF power (W)	DC power (W)	Ar flow rate (sccm)	Pressure (mbar)	T-T (mm)	T-S (mm)	Deposition time (sec)
1	300	300	10	1.33E-3	100	190	4,312
	300	300	30	3.99E-3	100	190	6,320
	300	300	50	6.65E-3	100	190	7,621
2	0	600	10	1.33E-3	100	190	4,129
	0	600	30	3.99E-3	100	190	4,747

Based on the process parameters from Table 6.13, OLEDs were fabricated and measured for evaluations. *I-V-L* characteristics of five OLEDs arbitrarily taken out of each process are shown in Figure 6.29.

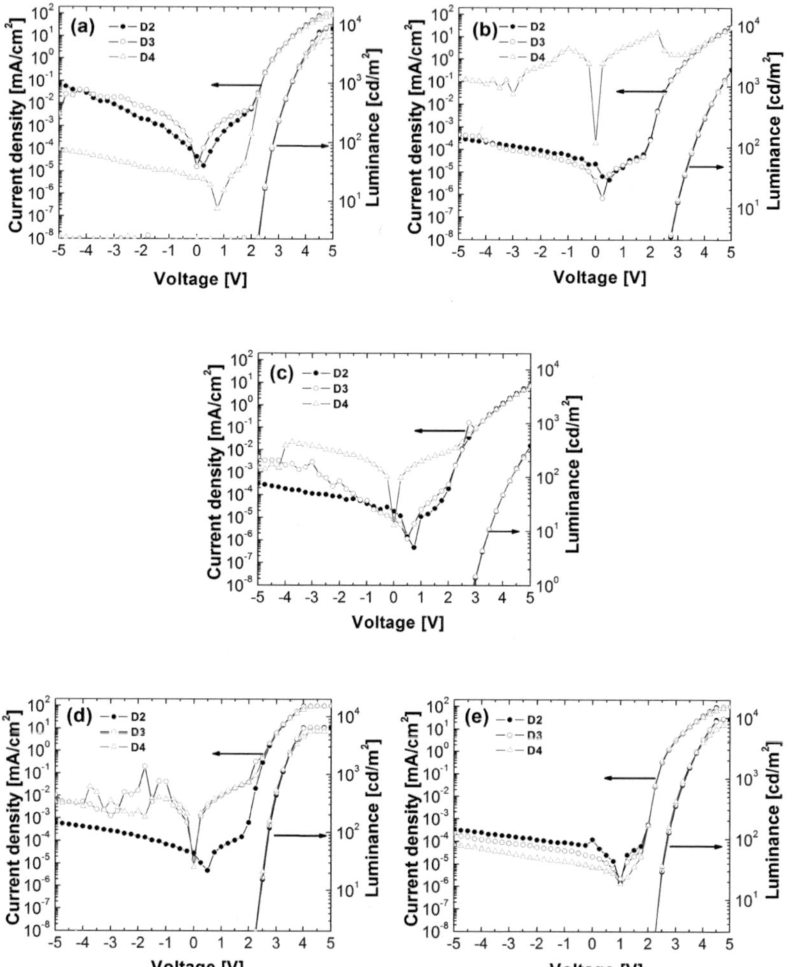

Figure 6.29: *I-V-L* characteristics of OLEDs. Figure (a), (b) and (c) are the results from the OLED processed at a RF portion of 50% and a pressure of 1.33, 3.99 and 6.65×10^{-3} mbar, respectively. OLED (d) and (e) are processed at a RF portion of 0% and a pressure of 1.33 and 3.99×10^{-3} mbar.

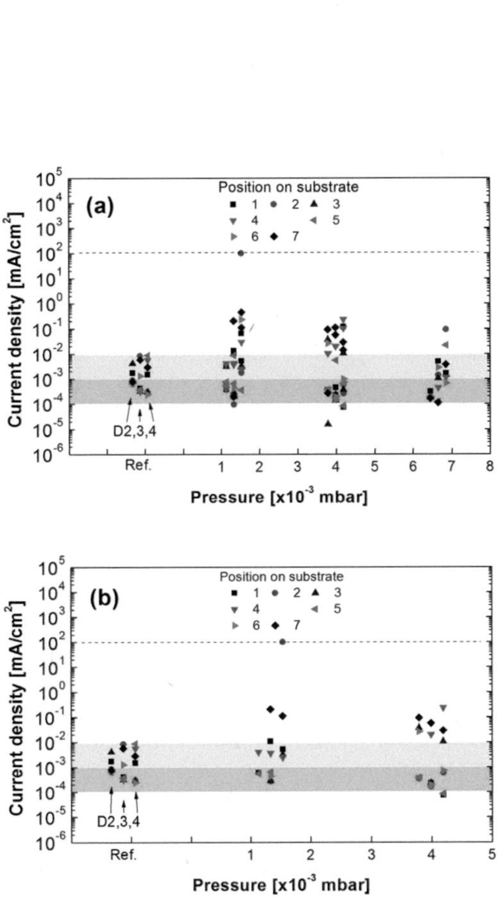

Figure 6.30: Collected leakage current values at -5V from the reference OLED and the OLEDs having sputtered cathode. Figure (a) and (b) show the leakage current densities from the processes with a RF portion of 50% and 0 %, respectively.

The values of the leakage current densities are collected and indicated in Figure 6.30. It was comparatively weak but it could be seen that the overall leakage current tends to decreases as the process pressure increases. However, Figure 6.30 (b) did not show even such a weak change. Interestingly, all D4 OLEDs of the position 2 at the pressure of 1.33×10^{-3} mbar reached the current compliance level of 100 mA/cm^2. Except for these positions, no obvious position-dependant trend was observed, but it should be noted that higher process pressure and lower discharge voltage are able to suppress the leakage current to some extent. However, as seen in the experiments with power mixing ratio, an increase of deposition time let the organic layers exposed longer to the plasma radiation, which can result in a decrease of forward current.

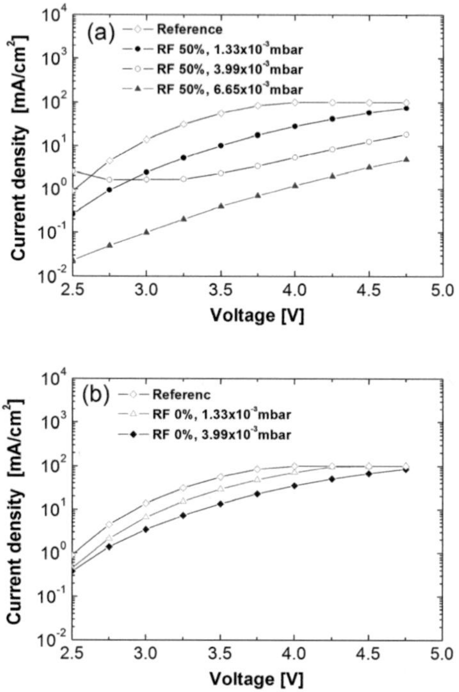

Figure 6.31: Forward current densities of OLEDs processed at different process pressures. Figure (a) and (b) are from the process with a RF portion of 50% and 0 %, respectively.

The forward *I-V* characteristics of the OLEDs are shown in Figure 6.31. The reference OLED exhibited the high current density of 100 mA/cm^2 at 4 Volts, but the other OLEDs having sputtered Al cathodes showed lower current densities than the reference OLED. This is obviously the effect of plasma radiation alike the experiments with power mixing ratio. As the process pressure increased, the deposition rate decreased due to the collision. In case of the RF portion of 50%, the resulting process time was prolonged from 4,312 s at the process pressure of 1.33×10^{-3} mbar to 7,621 s at the process pressure of 6.65×10^{-3} mbar. This led to a higher radiant energy. Overall current densities with the Al cathode deposited at the RF portion of 0% are higher than with that of the RF portion of 50%. This is because of the lower radiation intensity and shorter process time.

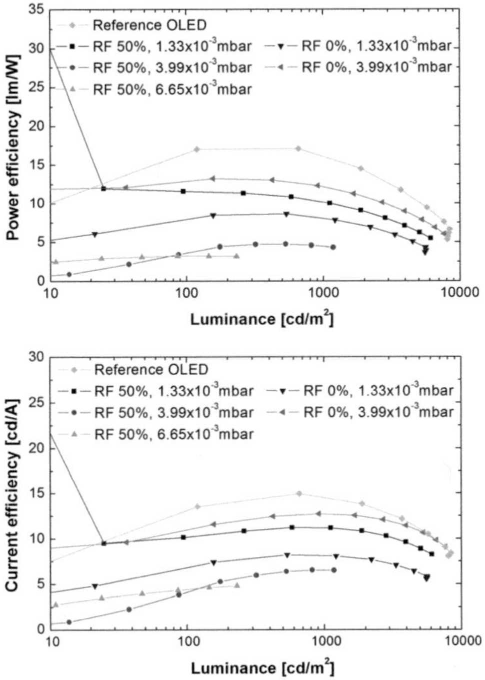

Figure 6.32: Power and current efficiencies of OLEDs.

Power and current efficiencies of the OLEDs are shown in Figure 6.32. Most of efficiencies are showing the anticipated trend, except for the RF 0% at 1.33×10^{-3} mbar. The reason can not be explained with the charge carrier imbalance in the emission zone because the forward current was the highest. It may stand for that the efficiency of OLED was influenced by the structural or optical properties at the metal/organic interface as well as the charge balance.

Table 6.14. Power and current efficiencies of OLEDs with respect to the luminance of 100 and 1,000 cd/m^2.

Process parameter	PE at 100 cd/m^2	PE at 1,000 cd/m^2	CE at 100 cd/m^2	CE at 1,000 cd/m^2
Reference	16.57	16.15	13.18	14.49
RF 50% at 1.33×10^{-3}	11.54	10.26	10.22	11.18
RF 50% at 3.99×10^{-3}	3.65	4.34	4.14	6.56
RF 50% at 6.65×10^{-3}	3.26	-	4.47	-
RF 0% at 1.33×10^{-3}	7.86	8.02	6.84	8.14
RF 0% at 3.99×10^{-3}	12.86	12.10	11.01	12.65

Six 4Q substrates from each process were taken for lifetime measurements, and three substrates were tested at a same current density. 20 and 30 mA/cm^2 was set to the OLEDs, and the evaluated lifetimes and the sudden death percentages during the lifetime measurement are indicated in Table 6.15. Overall, lower process pressure yielded lower driving voltage and longer lifetime at both powers. However, a higher sudden death rate was found at lower process pressure due to the presence of higher leakage current in some OLEDs. In particular, the OLEDs with Al cathode deposited at the RF portion of 0% and the pressure of 1.33×10^{-3} mbar were destroyed completely at 30 mA/cm^2. Compared with the results processed at the RF portion of 50% at the same pressure, the sudden death ratio shows that an Al sputter deposition by a lower sputter voltage is better at a low process pressure for a stable operation of OLED in spite of higher driving voltage. However, at an adequately high process pressure, a low plasma density, despite the high discharge voltage is beneficial to yield a comparable stability with higher OLED efficiency. This was obviously seen from extrapolated lifetime with respect to luminance in Table 6.16. However, the sudden death problem, the higher driving voltage and the lower efficiency than the reference OLED having evaporated electrode still remain as negative aspects in using FTS system for electrode deposition on organic layer.

Table 6.15. Measured lifetimes of OLEDs

Process parameter	Voltage (V)		Average lifetime (hour)		Sudden death (%)	
	at 20 mA/cm^2	at 30 mA/cm^2	at 20 mA/cm^2	at 30 mA/cm^2	at 20 mA/cm^2	at 30 mA/cm^2
Reference	3.1	3.2	3,583	1,758	0	0
RF 50% at 1.33×10^{-3}	3.80	4.00	1,362	463	16.67	33.33
RF 50% at 3.99×10^{-3}	4.38	4.60	1,375	516	0	16.67
RF 50% at 6.65×10^{-3}	5.24	5.53	990	480	0	16.67
RF 0% at 1.33×10^{-3}	3.31	3.51	2,437	536	50	100
RF 0% at 3.99×10^{-3}	3.70	3.90	1,309	483	25	0

Table 6.16. Extrapolated lifetime with respect to the luminance

Process parameter	Lifetime (hrs)	
	at 500 cd/m^2	at 1000 cd/m^2
RF 50% at 1.33×10^{-3}	22,423	5,391
RF 50% at 3.99×10^{-3}	11,704	2,947
RF 50% at 6.65×10^{-3}	3,342	1,151
RF 0% at 1.33×10^{-3}	5,766	2,560
RF 0% at 3.99×10^{-3}	43,492	8,488

6.1.4 Al deposition on organic layers using dual magnetron sputtering (DMS)

In this section, the characteristics of OLED having an Al cathode deposited by DMS are reported. DMS with a bipolar switching unit was introduced in section 4.3.3 with the schematic diagram of the realized DMS hardware configuration in the Inline equipment. One particular point in this experiment was that the moving carrier was not driven several times in front of the target, but only one time to deposit a film for a required thickness. The process parameters are listed in Table 6.17. In order to induce a frequent collision of Al atoms with background gases, Ar 500 sccm was injected to the chamber for an Ar process pressure of 5×10^{-3} hPa. 500 W DC power was applied to both cathodes alternately, and the carrier moving speed was adjusted at 0.14 m/min for a 100 nm Al layer in order to compare the characteristics of OLED with the previous experiments. The notable difference was that the present configuration has yielded a high deposition rate which was nearly fourteen times higher than the deposition rate obtained from the experiment with the normal DC power and the anode shield, as listed in Table 6.2. It means that the anode shield installed for the homogeneity of the deposited film screens a significant amount of sputtered particles. Furthermore, the DC voltage and DC current were also different from those of the normal sputtering under the same process parameters. Especially the DC current was increased with the DMS configuration, resulting in a lower DC voltage. The voltage and current are, however, root-mean-square (RMS) values from the output of the power supply. Generally, the peak voltage of the pulsed DC power is much higher than that of the normal DC power, and it is created by power supply to promote plasma re-establishment [168]. Therefore, the voltages for sputtering in both power supplies are almost similar with each other.

Table 6.17. Process parameters for Al deposition using DMS

DC power (W)	Ar flow (sccm)	Pressure (hPa)	Carrier speed (m/min)	Deposition rate (nm/speed.oscillation)	DC voltage (V)	DC current (A)
500	500	5×10^{-3}	0.14	714.28	200	2.5

For experiments, five 4Q substrates were used alike the former experiments. In this experiment, p-i-n phosphorescent green OLEDs were fabricated. The organic stack was indicated in Table 5.2. *I-V-L* characteristics of the fabricated OLEDs were measured by the Autronic Melcher, and the results of leakage current, forward current and efficiency from the *I-V-L* characteristics were compared with those of the reference OLED having evaporated Yb cathode.

I-V-L characteristics of two 4Q substrates arbitrarily selected are shown in Figure 6.33. Two substrates showed very low leakage current densities ranging within 10^{-3} and 10^{-4} mA/cm^2 and normal forward current densities of p-i-n OLED. The resulting luminance of the reference OLED was lower than of the OLED having Al cathode. For a better comparison of OLED performance, *I-V-L* and efficiency curves of D4 OLEDs out of the two substrates are displayed together in Figure 6.34. The two *I-V-L* curves are almost identical, but the efficiencies are different. The efficiency of the OLED having Al cathode was much higher than the reference OLED. This was due to the higher reflectance of the Al cathode and the better coupling of the reflected light within the OLED. OLED parameters are listed in Table 6.18. The driving voltage of the OLED having sputtered Al cathode became lower than the reference OLED, and the efficiencies were nearly double to those of the reference OLED.

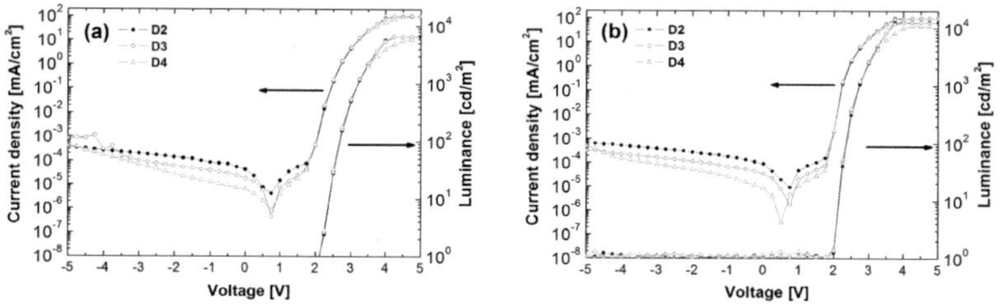

Figure 6.33: *I-V-L* characteristics of fabricated OLEDs; (a) reference OLEDs with Yb cathode and (b) OLEDs with sputtered Al cathode.

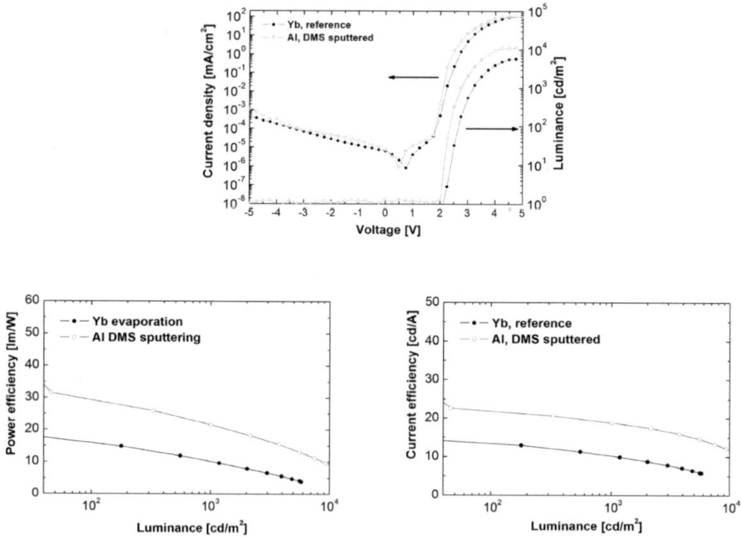

Figure 6.34: Comparisons of I-V-L characteristics and efficiencies of OLEDs. The OLED having sputtered Al cathode exhibits higher efficiency due to higher reflectance of Al than Yb and a resulting better light coupling from the OLED.

Table 6.18. OLED parameters from the fabricated OLEDs

Cathode	at 100 cd/m²			at 1,000 cd/m²		
	V [V]	PE [lm/W]	CE [cd/A]	V [V]	PE [lm/W]	CE [cd/A]
Yb, reference	2.77	15.85	13.23	3.22	10.07	10.00
Al, sputtered	2.35	29.37	21.33	2.76	21.57	18.60

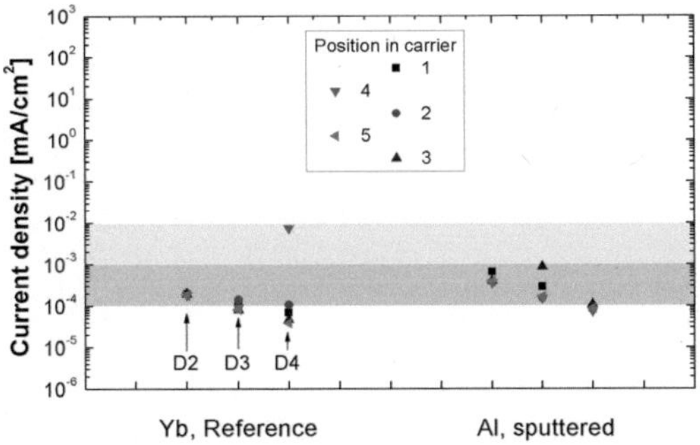

Figure 6.35: Collected leakage current densities of OLEDs. Both OLEDs exhibit low leakage current levels under 1×10^{-3} mA/cm².

All the leakage current densities were collected and indicated in Figure 6.35. As one can see, the OLEDs with Al cathode showed leakage current values as low as those of the reference OLEDs. If these leakage currents are compared with the leakage current of which the Al cathode was deposited at the same process parameters with the original hardware configuration, the obtained result was very encouraging in reducing the penetration of the Al atoms.

The cause may be simply ascribed to the low discharge voltage, but it was already shown in Section 6.1.2 that the leakage currents were still high at the discharge voltage of 214 V. A realistic cause can be inferred from the removal of the anode shields and the corresponding wider transport angle of the sputtered Al flux, as illustrated in Figure 5.18. In contrast with the T-S distance of 84.1 mm of the original hardware configuration, the DMS configuration might provide a much longer distance for the sputtered Al atoms. At a properly high process pressure, such as 5×10^{-3} hPa, the sputtered Al atoms could collide more often with background gases due to the given longer transport path, being thermalized. The Al atoms traveling much farther toward the carrier could be then condensed on the substrate with low incident energy before the substrate entered an intensive deposition region, which is namely the front of the sputter cathode. The average energy of the sputtered Al atoms is estimated to be much higher there due to shorter transport path length. It is important to remind that the carrier moved slowly and passed the two sputter cathodes only once. The pre-deposited Al film could act as a protective layer against a further energy transfer in the intensive deposition region. By means of the single pass of the carrier, a low damage sputter deposition of Al cathode could be possible without a significant increase of leakage currents. One may compare this result with that of the experiments by the FTS. The T-S distance at the FTS experiments was fixed at 190 mm, and it was also a quite far distance. However, if one calculates the energy transfer function ζ_L at different incident angles according to Eq. (4. 1. 12), it can be found that the energy transfer function becomes lower by an oblique projection of sputtered Al atoms in the DMS configuration than mostly 0° on average in the FTS system.

Not only the low leakage current, a high forward current was also observed, which was comparable to those of the reference OLED. Possible reasons are a screening of the plasma radiation by the pre-deposited Al layer and the radiation zone shift. The pre-deposited Al layer could protect the surface of the organic layer from the plasma radiation. Especially, Al is a good reflector in the UV-Vis-IR range, and a sufficiently thick Al layer might be deposited before the radiation acts seriously on the metal/organic interface. By the help of the bipolar unit and dual magnetrons, the plasma was also formed between the two sputter cathodes, being slightly shifted towards the other sputter target nearby. As shown in Figure 4.17, the plasma spread not only forward to the substrate, but also to the other sputter cathode by a potential difference and a magnetic field. Especially some of the secondary electrons can be guided along the magnetic field formed between the two magnetrons.

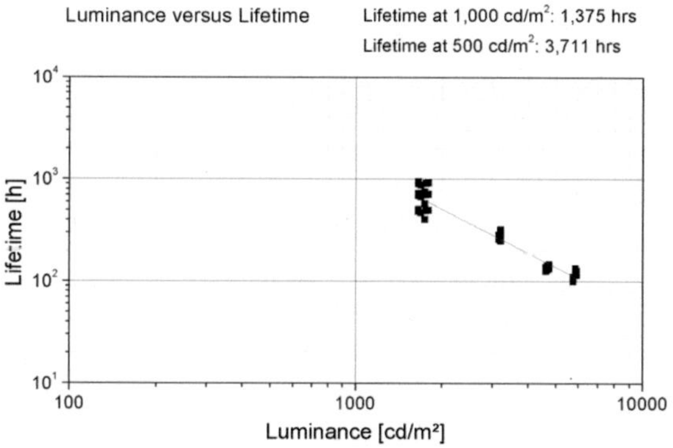

Figure 6.36: Extrapolated lifetime of OLED having sputtered Al cathode.

In order to examine the stability of the OLEDs, lifetime measurements were carried out. Four 4Q substrates were tested in BOTEST tool at the current density of 10, 20, 30, and 40 mA/cm^2 for the evaluation of extrapolated lifetime with respect to luminance. The result is shown in Figure 6.36. The OLEDs did not have the sudden death problem at all, but the extrapolated lifetime was quite shorter than the p-i-n phosphorescent orange OLEDs fabricated in cluster type equipment. It can be simply attributed to the characteristic of equipment or the chemical stability of emitter materials, but the lifetime was nevertheless short. Another effect such as the luminance decay depending on blocking material can be introduced [169], but no further investigations were performed in this study.

6.1.5 Summary and conclusion

OLEDs having Al top electrodes deposited by DC magnetron sputtering were demonstrated. The Al sputtering deposition on organic layers resulted in large leakage currents and lower forward current in the fabricated OLEDs. The OLEDs possessing the larger leakage current showed bright spots on the active area and shorter lifetime, so various experiments were performed to identify the origin of the degradations. In parallel, several attempts were carried out to prove whether previous studies are satisfactory to understand the results. The optical constants obtained from organic layers by ellipsometry showed no change in the optical band gap, which has been a dominant explanation during the last decade. Based on the operating principle of magnetron sputtering, it was found that the high energy of sputtered Al atoms was the dominant source responsible for the leakage current. The highly energetic Al atoms could deeply penetrate into the organic layers by a collision cascade within the layer from a continuous energy transfer. In order to prove the metal penetration, LDI-TOF-MS was used. The results showed the obvious existence of Al atoms inside the organic layers, and it assured that the high leakage current was induced by conductive localized paths provided by Al metal particles.

The plasma radiation effect on the metal/organic layer interface was also investigated. An increase of process pressure resulted in longer deposition rate and longer exposure time of the organic layers to the plasma radiation. As a result, the forward current densities of the OLEDs decreased with the increasing process pressure, and it reflects that the radiative particles out of plasma influenced the metal/organic layer interface.

In order to reduce the leakage current, RF superimposed DC power was applied to decrease the DC voltage, which is thought to be the dominant parameters to control the energy of particles. A lower voltage down to 214 V was achieved by the addition of high RF power. However, the leakage current remained unchanged, and a decrease of forward current was also found. Hence an improvement of degradation was not accomplished by the RF superimposed DC sputtering.

As another attempt to reduce the leakage current, Al cathodes were deposited by means of a FTS system with various process parameters. Through many experiments, the FTS system was evaluated whether it is a suitable method for damage-free sputtering of Al top electrode deposition on large area substrates.

In the experiments with various mixing ratio of RF and DC power, the best mixing ratio was found at the RF portion of 16.67% which resulted in the lowest tolerable leakage current and longest lifetime. However, this parameter also could not avoid an effect of plasma radiation, and driving voltage was increased due to lower forward current resulting from the radiation. This radiation turned out to be mostly a VUV emission from plasma based on the operation mechanisms of normal magnetron sputtering and FTS. As the RF portion in total RF superimposed DC power increased, the DC voltage decreased by the increase of plasma density. This effect resulted in a lower sputter yield and deposition rate, and the process time to deposit a certain thickness of electrode was prolonged. Radiant energy was calculated in terms of an approximation of ion density from the FTS cathode *I-V* characteristics. The values were in good agreement with an inverse trend of forward current densities of OLEDs. Especially, at the RF portion over 50%, plasma mode changed from DC mode to AC mode. AC mode sputtering gave rise to an ion bombardment onto the substrate, resulting in the additional decrease of forward current and the highest leakage current level. Eventually, the decrease of the forward current resulted in an imbalance of charge carriers in emission zone, and the efficiency of OLEDs decreased dramatically due to a deficiency of electrons. As the result, the extrapolated lifetime with respect to the luminance of the OLEDs having sputtered Al cathode could not reach the lifetime of the reference OLED.

In experiments with various process pressures, leakage currents tended to decrease as the process pressure increased. This was the same effect which was observed in the section using normal sputtering system. Reduced energy of sputtered Al atoms by collisions with background gases was successfully introduced in explaining the results. This effect seemed to be a better match at a low discharge voltage, but the resulting lower deposition rate and the longer process time resulted in more degradation of the metal/organic interface, leading to a higher driving voltage of the OLEDs. In case of lifetime, OLEDs processed at lower pressure exhibited a longer lifetime on average, but most of all OLEDs experienced a sudden death problem during the measurement.

By using the FTS system and RF superimposed DC power, the leakage current was expected to be decreased with the best mixing ratio of RF and DC power and high process pressure. However, it resulted in a dramatic decrease of the deposition rate, and the organic layers were exposed to plasma for a longer period. As the result, it was also difficult for the FTS system to yield a process that can satisfy both leakage current and forward current requirements. Furthermore, very long deposition rate like several thousand seconds is apparently a vast obstacle that hinders its application to OLED mass production, where three or four minutes tact time is required.

A successful result was obtained at last by means of a bipolar pulse unit and a dual magnetron sputtering in the Inline equipment. A hardware modification was also carried out to maximize the merits of the bipolar pulse dual magnetron sputtering. Process parameters, in particular, carrier moving sequence and process pressure were carefully selected. As a result, the deposition rate increased dramatically, and the leakage current was successfully suppressed without losing the forward current. Comparing with the reference OLED having evaporated Yb cathode, the performance of OLED fabricated with sputtered Al cathode was superior in driving voltage and efficiencies. Especially the efficiency of the OLED was almost double to that of the reference OLED with the help of the higher reflectance of Al film. The reason for the results was explained with a wider transport of sputtered Al atoms and corresponding lower energy transfer function at a wider angle.

6.2 Transparent cathode deposition on OLED using facing target sputtering

In section 4.5, the results from literature regarding ITO top electrode deposition have been discussed. Typically, the ITO deposition directly on organic layers also resulted in serious damage. In this study, a thin Ag was thus deposited first on organic layer for a good conductivity and a buffer layer before the ITO deposition in order to reduce the sputtering damage. First of all, phenomena related with damage from the ITO sputtering were investigated in order to find out factors which should be eliminated for a low damage ITO deposition. For ITO deposition, the process parameters from the Al depositions in section 6.1.3 were identically applied to compare the film properties of the two different materials and to analyze the deposition mechanism. In OLED fabrication, many trials by varying the thickness of the organic layers and the Ag cathode were also performed to reduce the degradation of OLED. Based on the results, ITO layers were deposited with two steps to achieve a low damage sputtering and a low resistive ITO layer at the same time. All of the fabricated OLEDs were evaluated by typical *I-V-L* and lifetime measurement alike the experiments for Al deposition.

6.2.1 ITO deposition on organic layers

As a preliminary work, ITO deposition on an organic and Ag layer was performed. On a precleaned glass substrate, 50 nm thick NET5:NDN1 (Novaled AG) was deposited as a test organic layer that has been implemented as ETL in the former experiments. A 20 nm Ag layer was subsequently deposited on the organic layer by thermal evaporation through another shadow mask in order to distinguish each deposition area. An arbitrarily selected process parameter for 100 nm thick ITO layer is listed in Table 6.19. This parameter yielded high transparency (>80% in visible range) and low sheet resistance (45.09 Ω/□). The ITO layer was deposited on the entire surface without shadow mask.

Table 6.19. Process parameter for ITO deposition

RF power (W)	DC power (W)	Ar flow rate (sccm)	O_2 flow rate (sccm)	Pressure (mbar)	T-T (mm)	T-S (mm)	Deposition time (sec)
500	500	10	0.1	1.33×10^{-3}	100	190	2,556

6.2.1.1 Observation of degradation

The films deposited on the glass substrate were observed by microscope (SMZ 1500, Nikon Co. Ltd.), and the images are shown in Figure 6.37. The Glass/NET5:NDN1/Ag/ITO and Glass/NET5:NDN1/ITO structures are distinguished by colors. Interestingly, net-like structures were found in both structures. A part of the ITO film was stripped by Kapton band, and the surface images including an unpeeled region are shown in Figure 6.37 (c) and (d). Certain patterns which remained on the glass substrate along the same pattern of the ITO layer were visible.

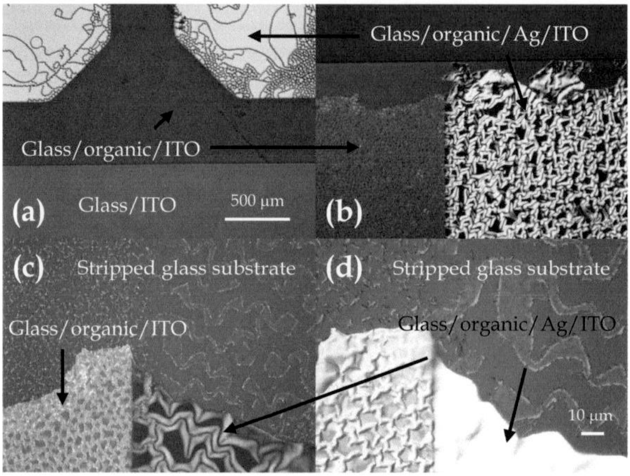

Figure 6.37: Microscopic images of degradation in Glass/NET5:NDN1/Ag/ITO structure. The images from (a) to (d) are observed with gradually increasing magnification.

Since only a planar view is possible by the microscope, a high-resolution SEM was used to observe the net-like structure microscopically. Four SEM images are demonstrated in Figure 6.38. Sinusoidal buckling waves are clearly visible in the images, which are commonly found in stress-relief phenomena and resulting delamination in multilayer depositions [170-172]. It is a quite undesirable result in depositing an ITO layer on organic and Ag layer by magnetron sputtering, and is believed to originate from the residual stress of the ITO film.

As discussed in Section 5.4.9, the thermal contribution results from the difference of the thermal expansion coefficients of films and substrates during the film deposition and the stress measurement. The latter case can be neglected when the stress is measured at room temperature, but the former case can be considerable in high-temperature deposition processes. Since the thermal expansion coefficient of NET5:NDN1 is unknown and the ITO deposition was performed at room temperature, the former case may not be so considerable. The intrinsic stress is, however, built up in the films during magnetron sputtering deposition even at moderate temperatures. In this case, an 'atomic peening' model is appropriate to describe the phenomena. It leads to an intrinsic stress in films prepared by sputtering or ion-assisted deposition techniques involving energetic particles bombardment onto the growing films.

Figure 6.38: SEM images of Glass/organic/Ag/ITO structure with increasing magnitude. The net-like structure turned out to be a buckled ITO, forming hollow tunnels, and it seems to be a kind of a pattern that the stress of the ITO film was released. Some organic materials and Ag remain under the ITO film and on the glass substrate.

A compressive intrinsic stress is usually found in films when energetic particles contribute to film deposition or bombardment on the growing film. In a FTS system, various types of atoms and molecules out of ITO target as well as recoiled Ar neutrals and ionized oxygen are those particles. Since the film was expanded in opposite direction from the substrate, it is evidently compressive stress. A tensile stress is induced during the deposition of films from very low energetic particles condensed on the substrate surface. A high level of the tensile stress can crack the film and substrate while the compressive stress leads to a delamination of film from the substrate surface. Therefore, it assures that the stress was compressive. The extrinsic contribution was induced by external factors subsequent to the deposition of films such as adsorption of water vapor through a porous film. However, the buckling waves are probably not due to the extrinsic contribution because such phenomena were observed as soon as the sample was evacuated from the vacuum chamber and subsequently transported to nitrogen-filled glove box.

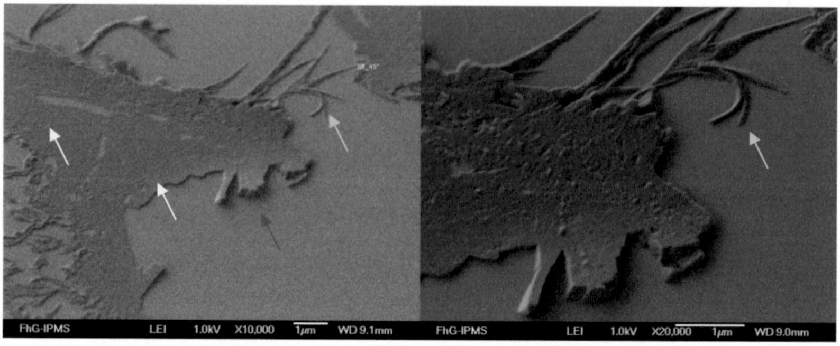

Figure 6.39: SEM images of remaining organic/Ag stack. The specimen was observed 45° tilted, and it is obviously seen that the organic/Ag stack was seriously degraded. It is distinguished by very thin areas (white arrows) and thick recrystallized areas (red arrows), and some indication of stretched recrystallization processes (green arrows).

The remaining organic/Ag stack was also observed. Figure 6.39 contains SEM images on the glass substrate after the ITO was striped. Some residual film was observed in the area where the ITO film did not buckle. It was still adhesive to the glass substrate, and some recrystallized structures appeared specifically. While the area indicated by white arrows seems to sustain an original organic/Ag film form, the other areas indicated by red and green are estimated as a result of a recrystallization of organic and Ag layer due to the high energy transfer from ITO sputtering. As discussed in section 4.4, ITO sputtering provides even in FTS system a more aggressive situation than Al sputtering due to energetic Ar neutrals and reactive gases like oxygen, to which most organic materials are vulnerable.

In order to investigate another factor which can result in such degradation, a 200 nm thick Ag film deposited by PK500 magnetron sputter cathode was treated by Ar + O_2 plasma. During ITO sputtering, oxygen ions are produced not only by ionization of injected oxygen gas, but also by dissociation of ITO molecules resulted from ion bombardment with high energy onto target surface and collision with ions in bulk plasma. An ICP-type plasma chamber in the cluster equipment was used for the plasma treatment. The Ag surface was exposed to the plasma for five minutes at 500 W RF power and Ar:O_2 flow rate = 30:10 sccm. The treated Ag film was transported to the glove box, and subsequently to a microscope in order to minimize extrinsic effects from, for example, water vapor. Interestingly, buckling waves were also observed on the Ag surface alike ITO deposition on organic/Ag layer. The photographs of the surface are shown in Figure 6.40. The bombardment of oxygen ion might oxidize the Ag surface and result in an AgO_x layer at the Ag surface. The formation of AgO_x in oxygen plasma is generally attributed to neutral oxygen radicals in the plasma [173], and such an oxide layer is supposed to be a highly dielectric layer with a high film stress. The surface oxide layer striped off spontaneously after a long exposure to ambient as shown in Figure 6.40 (b). Therefore, a plasma surface treatment on Ag film before and after organic layers must be performed with only inert gas like Ar for a cleaning purpose, excluding reactive gases such as oxygen or nitrogen.

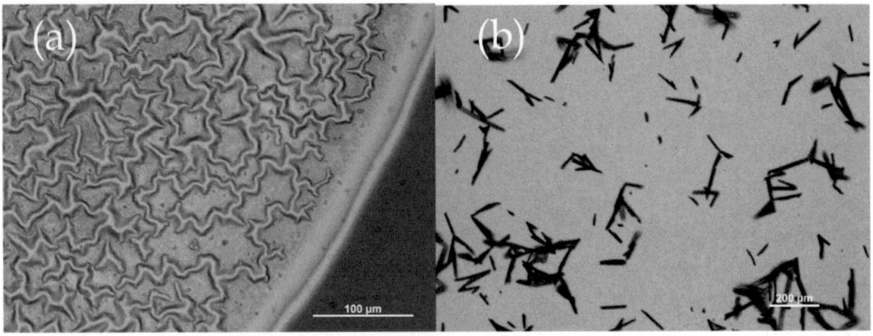

Figure 6.40: The surface of Ag after plasma treatment with $Ar+O_2$ gas mixture. Buckling waves were also observed on the Ag surface in figure (a). The oxidized film was finally peeled off spontaneously like figure (b).

An oxygen gas-free ITO process would be beneficial to reduce a considerable oxidation of thin Ag cathode. Additionally, investigations of the influence of deposition parameters on the residual stress level and the origin of the intrinsic stress developed in ITO films are essentially required in order to find ways to reduce and control the stress level which limits the adhesion property to an underlying layer and a poor performance of OLED, exactly, the injection of charge carriers. Not only the two requirements, the electrical and optical properties of ITO films, must be examined in detail, since they determine charge transport from the electrical contact to the entire OLED, and the optical output through the ITO layer. It should be identified as well whether a process parameter for a good quality ITO gives rise to less or higher damage in OLED. Therefore, the understanding of the ITO deposition and the influence of the process parameters are strongly necessary in order to accomplish a low damage ITO deposition.

6.2.2 Characteristics of ITO deposited by FTS system

The ITO films were prepared by a FTS system in the cluster equipment with RF superimposed DC power. The target to substrate (T-S) distance was 190 mm, and the target to target (T-T) distance was 100 mm. The vacuum chamber was evacuated down to a base pressure of about 2×10^{-7} mbar prior to every deposition. As explained above, all ITO films were deposited without oxygen gas in order to avoid a deterioration of underlying layers including Ag and organic.

Since the same process parameter sets from Al depositions were used for ITO depositions, one can understand the results and mechanism more easily by comparing them with the results of the Al depositions. The goal thickness of ITO films was again 100 nm. The ITO films were characterized by various instruments; UV-Vis spectrometer, four point probe, Hall measurement, ellipsometer, stress measurement, AFM, and X-ray diffraction (XRD).

In the first evaluation, the total power was fixed at 600 W and the RF portion was varied from 0 to 83.33 %. As seen in section 6.1.2 and 6.1.3.1, the advantage of the power mixture is that the DC cathode voltage can be altered by the mixing ratio. The process parameters and deposition time for a 100 nm thick ITO film are listed in Table 6.20. Note that the ITO deposition time at the same process parameter was much shorter than the Al deposition. It is difficult to define an exact sputtering yield of molecular mixtures, but the sputtering yields of In and Sn are respectively 2.135 and 1.569 which are higher than 0.992 of Al at an Ar incident energy of 500 eV [174]. One may estimate that the sputtering yield of a molecular solid composed of such atoms is possibly higher than Al in comparison with the results of Table 6.7.

Table 6.20. Process parameters for power mixing ratio variation

RF portion (%)	RF power (W)	DC power (W)	Ar flow rate (sccm)	Pressure (mbar)	T-T (mm)	T-S (mm)	Deposition time (sec)
0	0	600	30	3.99×10^{-3}	100	190	3,136
16.67	100	500	30	3.99×10^{-3}	100	190	3,259
50	300	300	30	3.99×10^{-3}	100	190	3,845
83.33	500	100	30	3.99×10^{-3}	100	190	6,469

Firstly, electrical characteristics of ITO films were investigated. The film resistivity ρ, carrier density n_e, and electron mobility μ_e were plotted together with the DC voltage in Figure 6.41. The DC voltage was decreased with increasing RF power. The DC voltage in ITO deposition appeared a bit lower than that in Al deposition, and is presumably due to a higher ionization rate of species which are produced during ITO sputtering. It is very interesting that the resistivity increased gradually with increasing RF portion and dropped dramatically at the RF portion of 83.33%. As seen in section 6.1.3.1, the region around the RF portion of 50 % turned out to be a transition point between DC and AC plasma excitation mode.

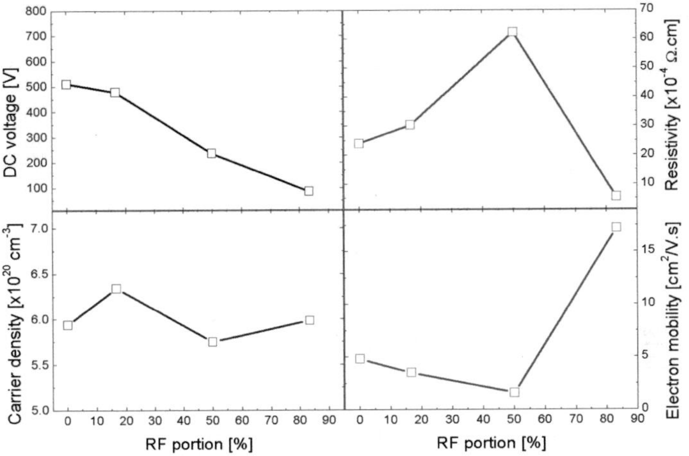

Figure 6.41: Electrical properties of ITO films. The variations of film resistivity ρ, carrier density n_e, and electron mobility μ_e with respect to the RF portion variation are shown with DC voltage.

In the DC operation mode up to a 50% RF portion, the increase of the film resistivity appeared to follow the decrease of the DC voltage well. This is a similar result to the Al deposition in DC operation mode, and it can be explained by either a decreasing deposition rate or a structural difference in the film due to a lower energy transfer by atomic and molecular particles resulted from the decreasing DC voltage. On the other hand, in the AC operation mode of the RF portion above 50%, the film resistivity abruptly exhibited a very low value of 5.4×10^{-4} μΩ·cm, which has not been reported in depositing ITO film at room temperature without additional oxygen gas. The electrical properties of ITO, mainly carrier density and mobility, are significantly affected by several factors which are strongly dependent on process parameters, doping concentration, oxygen content, etc.

The carrier density is taken into account first. It should be recalled that an ion bombardment exists in the AC operation mode as already shown in the section of Al depositions. Especially in an AC discharge with high frequency such as 13.56 MHz, reactive species with high density are created as atomic or excited oxygen and ozone forms [175]. Many researches reported that the increase of oxygen content resulted in a decrease of carrier density, which can be explained by an increasing degree of incorporation of interstitial oxygen [176, 177]. If more oxygen atoms were incorporated into the film through the ion bombardment effect, it might result in a decrease of carrier density. However, the carrier density shown in Figure 6.41 did not show a considerable change even at the high RF portion. That means that the incorporation of oxygen did not play a decisive role to change the resistivity. However, it was observed that the electron mobility dramatically increased at the RF portion above 50%.

Secondly, the structural properties of films were investigated by means of XRD. The measured XRD patterns are shown in Figure 6.42.

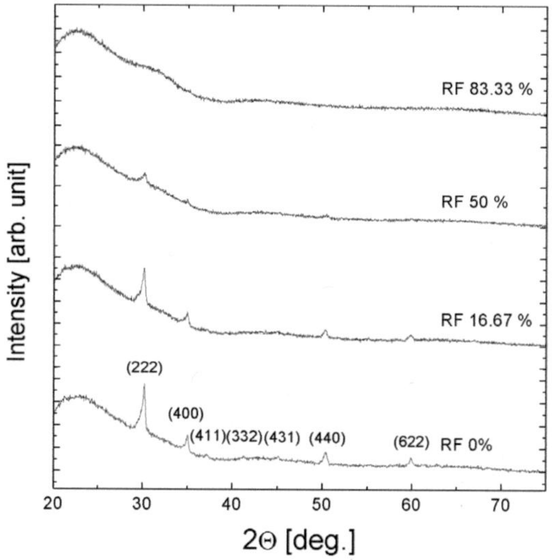

Figure 6.42: Comparison of XRD patterns of ITO films deposited at various RF power portions.

The wide peak at about 22° is due to the amorphous structure of the glass substrate. XRD patterns showed several peaks which correspond to crystal orientations of ITO. Among them, the (222), (400), (440), and (622) major diffraction peaks were apparently visible, and some broad peaks of (411), (332), and (431) were also recognized. Interestingly, the intensity of the major peaks decreased as the RF portion increased, and its trend seemed to be in accordance with that of the DC voltage. It means that the energy of particles generated during sputtering decreased with decreasing the DC voltage. The particles comprising sputtered atoms or molecules and recoiled Ar neutrals can possibly transfer less energy to the film being deposited, and this may result in a poor crystallinity of film and an almost amorphous film at last.

The film morphology was observed by AFM in order to compare the structural difference with the results of XRD. The AFM images are shown in Figure 6.43. Since the structure of ITO films seemed to follow the DC voltage, measured roughness values were plotted with the DC voltage in Figure 6.44.

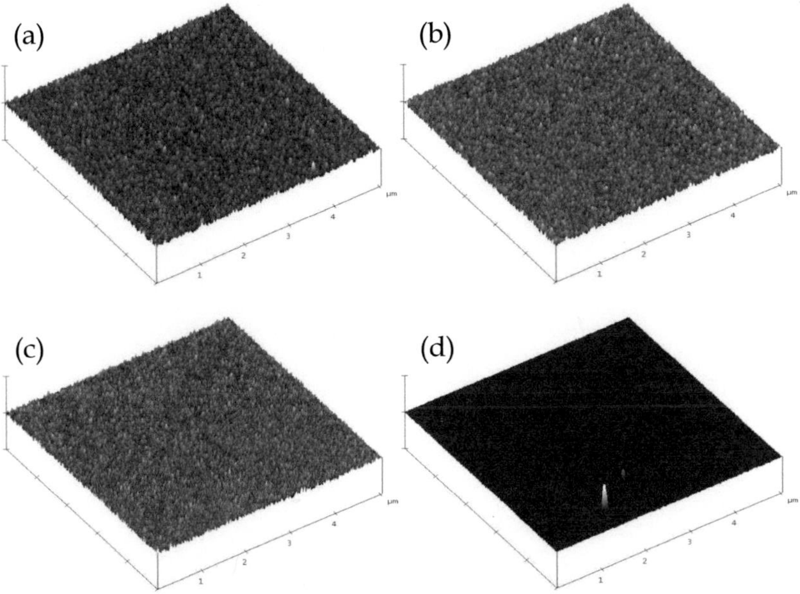

Figure 6.43: AFM images of ITO films displayed with y-scale of 50 nm. Measured area is 5 × 5 μm². The films were deposited at the RF portion of (a) 0%, (b) 16.65 %, (c) 50 % and (d) 83.33 %.

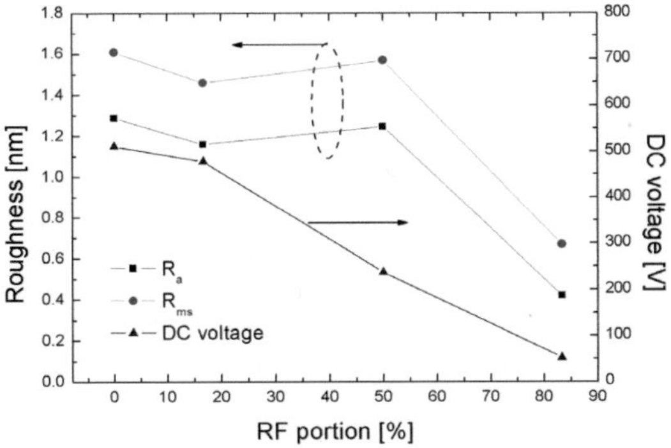

Figure 6.44: Plot of the measured roughness R_a and R_{ms} values with the DC voltage. The trend of the film roughness is in good agreement with DC voltage except for the film deposited at the RF portion of 50 %.

It is clearly seen in Figure 6.43 that the films deposited at the RF portion from 0 to 50 % are showing polycrystalline grain structures as observed in the XRD patterns. A pronounced difference among them is found from the morphology of the ITO film deposited at the high RF portion. The film in Figure 6.43 exhibited a very smooth surface which is considered as an amorphous film. In order to investigate the reason for the structural difference further, the effect of the DC voltage is taken into account. Figure 6.44 shows the roughness values and the DC voltage. The overall trend of the roughness values seemed to follow that of the DC voltage except for the ITO film deposited at the RF 50 %, and even the roughness rather increased. That means that the ITO films up to the RF 50 % grew as polycrystalline films in the DC excitation mode, and the dramatic decrease of the roughness at the RF portion 83.33% was possibly not influenced by only the DC voltage, but other factors, such as ion bombardment. The change of the plasma excitation mode was already introduced several times. There are literature reports regarding the deposition of TCOs assisted by an ion bombardment using an additional ion gun [178, 179]. The ion bombardment effect resulted in a crystallization of an amorphous film deposited and a preferred orientation appeared along the (222) orientation in those reports. However, the result of the ITO films of this study was completely in contrast with the previous reports, and a different energy and a high dose of bombarding ions might cause such difference. As shown in Figure 4.13, the average ion energy increases with increasing the RF portion. The ions in the bulk plasma fall onto the surface of the film during deposition with an energy corresponding to plasma potential relative to the substrate surface immersed. They can, however, sputter the condensed ITO film very less due to much lower energy than the ions from an ion gun, but make the film much denser. This densification effect gives a solution of a contradictory question that how an amorphous film can exhibit higher carrier mobility than a polycrystalline film.

The mobility term in Eq. (2. 2. 12) resulting from the ionized impurities is not taken into account because the carrier density was almost constant. The mobility due to lattice vibrations is also excluded because the film was deposited at room temperature and the measurement was performed also at room temperature. Hence one must investigate the influence of grain boundary scattering on the film mobility.

By the ion bombardment, the amorphous ITO film might be quite densified, so the resulting thickness of depletion layers between ITO molecules derived from Eq. (2. 2. 14) could be dramatically thinner. Therefore the carrier mobility could be greatly improved due to the lack of grain boundary scattering, such that the film could exhibit high carrier mobility in spite of the amorphous phase.

The film stress should be examined as well in order to avoid the buckling of the deposited ITO and underlying Ag layer. ITO films were deposited with the process parameters in Table 6.20. In Figure 6.45, the measured stress levels of the ITO films are indicated. The ITO films showed a compressive stress which one can see from the negative sign. Alike the other measurements, the ITO film showed a dramatic change at the RF portion of 83.33 %, while the film stress tended to decrease with increasing the RF portion up to 50 %. This result is attributed to the decrease of the DC voltage. It ensures the effect of the ion bombardment once more, since such highly compressive stress is generally observed where a considerable ion bombardment exists. However, the stress of the ITO films deposited by the FTS system was actually quite low compared with other studies. At a similar process condition, P. F. Carcia reported a highly compressive stress over one GPa, and T. C. Gorjanc reported 4.9 GPa [180, 181]. The lower stress obtained using FTS is undoubtedly due to the elimination of the impact of negatively charged oxygen, and FTS is estimated as a helpful method to reduce the sputtering-induced damage in OLED. One specific point is the low stress of ITO processed at the RF portion of 50 %, which was applied for the preliminary work in the former section, so the oxygen addition was the dominant reason responsible for the buckling of Ag/ITO structure. Detailed phenomena will be examined in the next section, the OLED fabrication part.

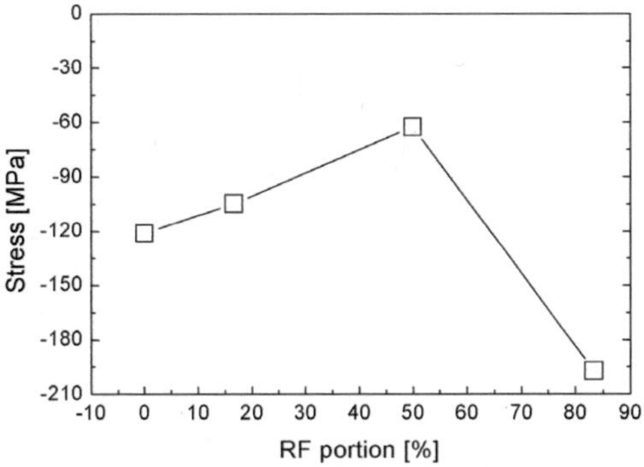

Figure 6.45: Measured ITO film stress on a Si wafer. The minus sign states a compressive stress. A highly compressive stress of around 197 MPa is found at the RF portion of 83.33% due to the bombardment of energetic ions in AC plasma mode.

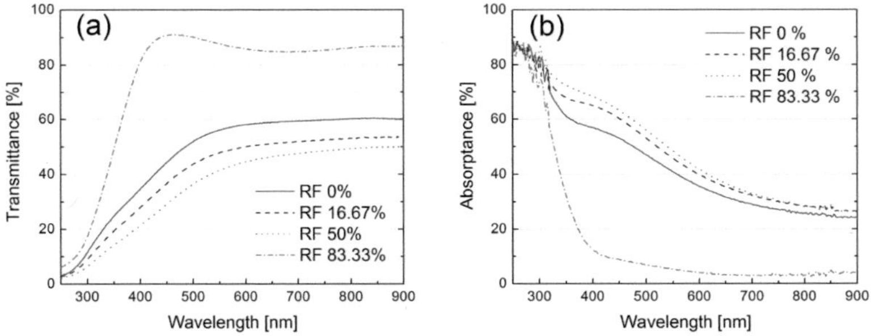

Figure 6.46: Transmittance T (a) and absorptance A curves (b) of ITO films deposited at various RF portions on glass substrate.

Optical characteristics such as transmittance, absorptance, and refractive index were investigated, since those properties are also important factors to determine the outcoupling efficiency of OLED. Transmittance and absorptance of ITO films deposited on corning eagle glass substrates were measured by UV-Vis spectrometer (Lambda 900, Perkin Elmer).

As shown in Figure 6.46, the optical characteristics of ITO films revealed a close relationship to the electrical and structural characteristics. The transmittance of ITO films decreased as the RF portion increased up to 50%, and the absorptance increases vice versa. From the previous measurements, the decrease of the carrier mobility turned out to be due to scattering at grain boundaries between polycrystallites. It was briefly discussed in section 2.2.1.3 that local deformations of the lattice can result in a tailing of the absorptance due to band-gap broadening or provide defect states between band gaps. Under process conditions where a less oriented polycrystalline film can grow, such high absorption can be observed in the film. It should be however noted here that the film may exhibit different characteristics when ITO films are deposited under an oxygen-mixed ambient. At the high RF portion of 83.33%, a highly transparent ITO film was obtained with very low absorption. The average transmittance T of the ITO film was derived as 87.29% including glass within the wavelength from 400 to 700 nm. This result is apparently a corresponding effect of improved film density with fewer defects due to ion bombardment. The defects in the film might be cured by ion bombardment with a moderate energy so that the band-gap could be observed clearly. An optical band-gap is then obtained by extrapolating the linear part of the absorptance curve to the wavelength axis. The intersection point with the wavelength was found at 389 nm, namely, 3.19 eV. This value is still much lower than those usually found from ITO films deposited at high temperature (~300 °C). Nevertheless, the characteristics of ITO films processed at room temperature without oxygen gas were excellent. However, one can already anticipate that the ion bombardment would damage the surface of the underlying organic layer.

The refractive index n was also measured by an ellipsometer. Depending on the index and the thickness of the dielectric medium, the interference of electromagnetic waves is influenced so significantly that a careful approach is needed to satisfy the electrical and optical properties of dielectric electrode simultaneously in OLED. In Figure 6.47, the measured refractive indices of ITO films are shown. Certainly, the refractive index n also satisfied the defect-induced scattering relation very well. Referring to Eq. (2. 2. 14) with $\varepsilon_r = n^2$ where the relative permeability μ_r is almost 1 for nonmagnetic material, one can find that the density of film, which can be inferred from the thickness of the depletion layer between grains, is directly dependent on a change of the optical property.

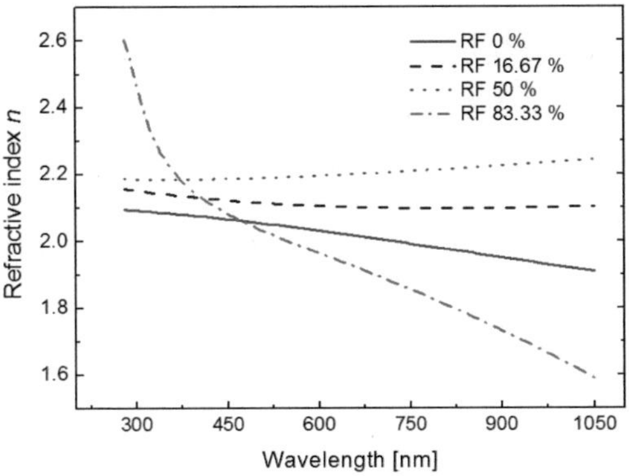

Figure 6.47: Refractive indices of ITO films deposited at different RF portions.

In the second evaluation, the process pressure was varied from 1 to 5×10^{-3} mbar at a constant process power as listed in Table 6.21. The process time was obtained for a 100 nm ITO layer. The process pressure influences the collision frequency between particles due to the change of the mean free path, and this effect is expected to alter the energy of particle generated during ITO sputtering alike the former experiments. Commonly seen is the increase of deposition time as the process pressure increases. As explained several times, this is due to the back-scattering of sputtered material by collision with gases.

Table 6.21. Process parameters for pressure variation

RF power (W)	DC power (W)	Ar flow rate (sccm)	Pressure (mbar)	T-T (mm)	T-S (mm)	Deposition time (sec)
300	300	10	1.33E-3	100	190	4,312
300	300	30	3.99E-3	100	190	6,320
300	300	50	6.65E-3	100	190	7,621

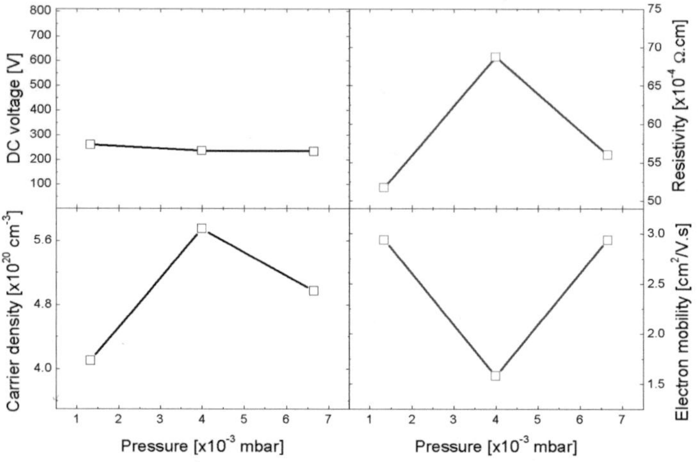

Figure 6.48: Electrical properties of ITO films. The variations of film resistivity ρ, carrier density n_e and electron mobility μ_e with respect to the pressure variation are indicated with DC voltage.

Identical measurements were carried out for the ITO characterizations. In Figure 6.48, electrical properties of ITO films deposited at different process pressures are shown. Since only a slight decrease of DC voltage was observed, the initial energy of particle before collision is assumed to be comparable. The lowest resistivity of the ITO film is found at the low pressure of 1.33×10^{-3} mbar, and this can be explained by a high energy of sputtered particles due to less collision and resulting high film density. By increasing process pressure, carrier density and mobility show a minimum at the process pressure of 3.99×10^{-3} mbar. This result however can not be explained by collision-induced energy loss mechanism. Thus, a structural property should be investigated as well.

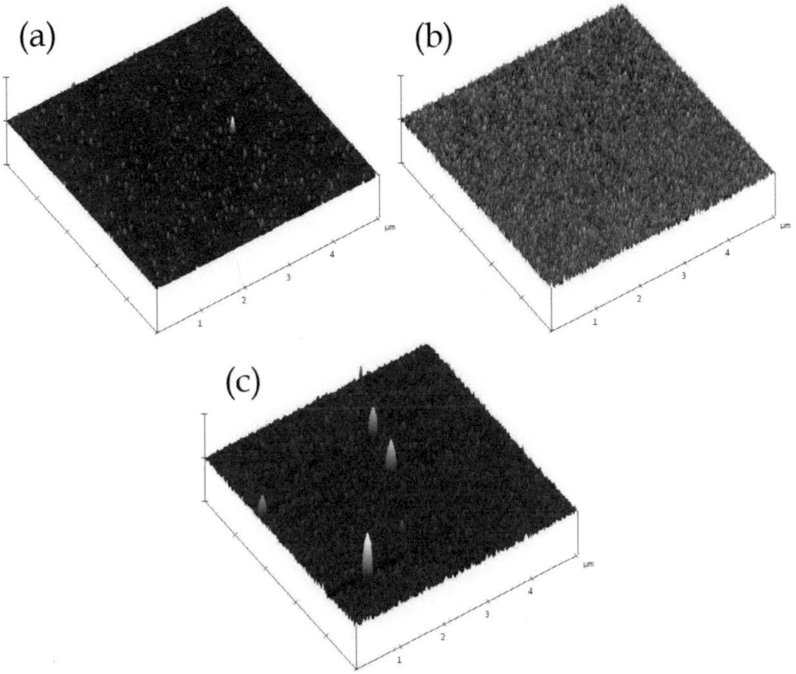

Figure 6.49: AFM images of ITO films deposited at different pressures, (a) 1.33×10^{-3} mbar, (b) 3.99×10^{-3} mbar and (c) 6.65×10^{-3} mbar.

The film morphology was observed by AFM, and the images are shown in Figure 6.49. The measured roughness values with DC voltage are indicated with respect to the process pressure in Figure 6.50. The ITO film in Figure 6.49 (a) deposited at 1.33×10^{-3} mbar had overall a smooth surface, but some grains were also observed. As discussed in the previous section, such a flat surface could result from the bombardment of energetic particles with high dose. Since the supplied power was the same for every process, the low process pressure could result in less collisions leading to a more effective bombardment with high energy. The roughness of the ITO film showed the minimum value among the ITO films as shown in Figure 6.50. At the process pressure of 3.99×10^{-3} mbar, a considerable amount of grains was visible. Increasing the process pressure further, some extremely large peaks were observed. They are supposed to be due to a particle contamination because their size was incomparably larger than the overall grain sizes. However, it is still difficult to analyze the reason for the transition of electrical properties as well as morphologies.

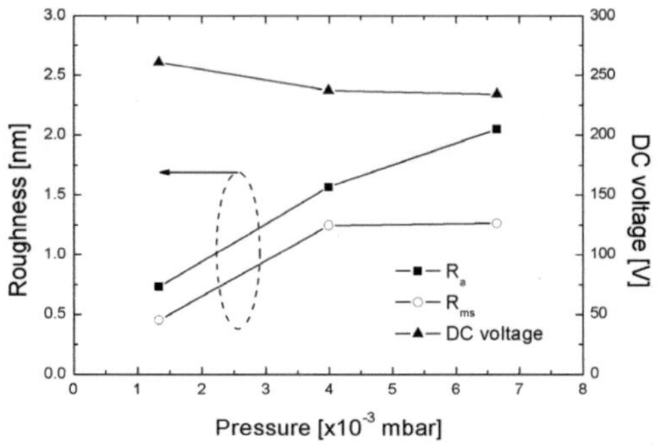

Figure 6.50: Plot of the measured roughness Ra and Rms values with the DC voltage from ITO films deposited at different process pressures.

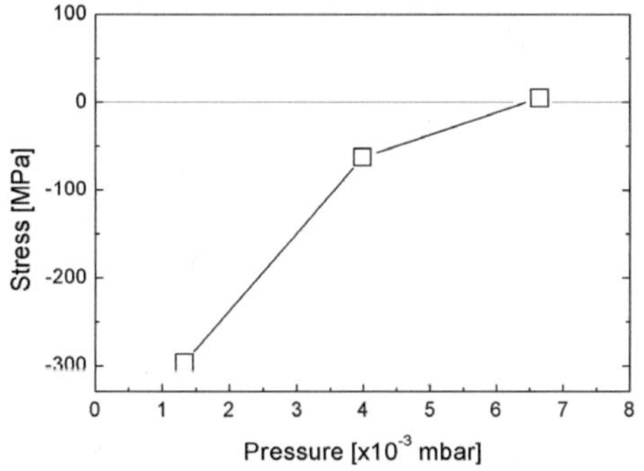

Figure 6.51: Measured ITO film stress on Si wafer. Very high compressive stress of 296.5 MPa is found when the ITO film is deposited at a low process pressure. With increasing the process pressure, a transition point of film stress from compressive to tensile is observed at around 6.3×10^{-3} mbar.

The results of stress measurement are shown in Figure 6.51. As expected, the ITO film deposited at the low process pressure of 1.33×10^{-3} mbar exhibits very high compressive stress due to significant bombardment. If this result is considered together with the process parameters of Table 6.19 for the ITO deposition test, which resulted in the serious buckling of the ITO film on an organic/Ag structure, one can estimate that the buckling can also occur by the high film stress of ITO as well as the deterioration of thin Ag layer due to oxygen ambient. As the process pressure increased further, the compressive stress decreased and eventually transformed to a tensile stress. The transition point was found around 6.3×10^{-3} mbar. The transition observed at the process pressure of 6.65×10^{-3} mbar in the electrical characteristics of the ITO films is probably caused by stress-related defects inside the film, and such defects might be cured by increasing the process pressure.

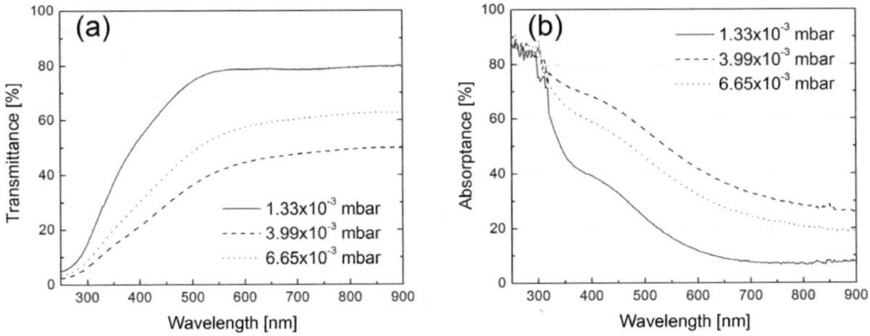

Figure 6.52: Transmittance (a) and absorptance (b) curves of ITO films deposited at different process pressures on glass substrate.

The optical characteristics of the ITO films corresponded to the electrical characteristics very well. In Figure 6.52, transmittance and absorptance curves of the ITO films deposited at different process pressures are shown. Overall tendencies were in good agreement with the change of resistivity comprising carrier density and mobility. The ITO film deposited at 1.33×10^{-3} mbar showed the highest transmittance, but the transmittance of the ITO film deposited at 6.65×10^{-3} mbar was also as high as over 50% in the range of interest.

Various characterizations of ITO films deposited by means of the FTS system have been performed in order to understand the behavior of ITO films depending on the process parameters. In particular, the ITO films were deposited without oxygen because the underlying thin Ag layer was seriously degraded exposing a buckling and complexes mixed with the organic layer. In ordinary ITO depositions on glass or rigid substrates, approaches to optimize ITO films for good electrical and optical properties are much easier, but such possibilities are greatly restricted due to the specific condition that the ITO films should be deposited on organic materials, which are extremely sensitive to highly energetic particles and oxygen. Therefore, several requirements must be satisfied in order to succeed a low-damage ITO sputtering on organic layer: low energy of particles generated by sputtering, good electrical properties as an electrode, low film stress for a less mechanical deterioration, high optical transparency for a better efficiency of OLED and commonly minimized oxygen addition. In fact, to fulfill all the requirements written above may be impossible. Nevertheless, many trials were carried out in order to reduce sputter damage and fabricate OLEDs having an ITO top electrode, and the results will be discussed in the next sections.

6.2.3 Characteristics of thin Ag layers on organic layers

As proposed in the previous section, a thin Ag layer was used as a highly conductive and buffer layer against sputter damage. Thin metal films have been used in semitransparent metal windows, optical filters, and modern optoelectronic devices like microcavity LEDs [182]. Optical and electrical properties of thin metals are, however, greatly altered by the disorder in the crystallinity of the metals. It has been known for several decades that very thin or discontinuous metal films show abnormal absorption and dispersion in the visible range, in contrast to sufficiently thick films or bulk metals [183-185]. Especially Ag, which exhibits the highest conductivity among metals, was also shown to have absorption and worse electrical conductivity for a very thin thickness (~15 nm) due to surface scattering. The electrical conductivity of the Ag layer is usually measured by four point probe. However, if the thickness is thinner than 10 nm, the conductivity exhibits abnormally low because the nuclei of condensed Ag atoms are not completely connected. The electrical characterization of thin Ag layers deposited on organic layers is thought to be more difficult, because the film morphology of organic thin films is usually poorer than glass or Si wafer. Therefore, a careful approach is needed for deposition with a precise deposition rate control and homogeneous film growth on a given substrate.

In order to observe the effect on the film morphology from different substrates, Ag layers were deposited directly on a Corning glass substrate and a 100 nm thick NET5 layer deposited on a Corning glass.

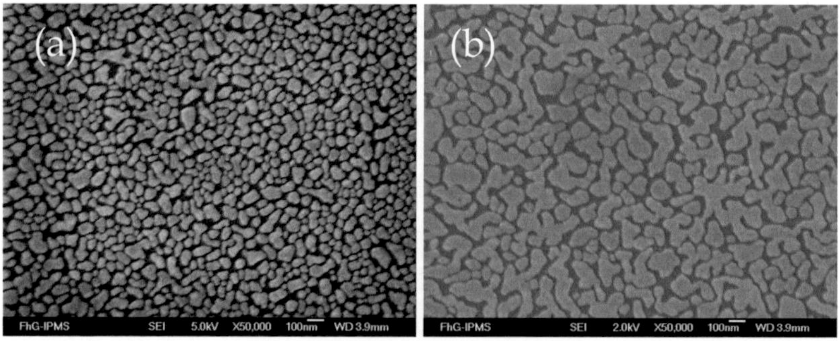

Figure 6.53: Film morphologies observed by SEM. Ag layers were deposited on glass substrate at the deposition rate of 0.3 Å/sec. The thickness of Ag was (a) 15 nm and (b) 20 nm. The images reprinted from Ref. [186].

In Figure 6.53, two SEM images of Ag layers deposited on glass substrate are shown. The thickness of the Ag layers was 15 and 20 nm, respectively. The 15 nm thick Ag layer still showed discontinuous grains. However, the grains of the 20 nm thick Ag layer began to get connected forming bound grains, which can transport electrical current. The sheet resistance of the thin Ag layer was measured by the four point probe. Since the four point probe method is quite destructive, the Ag layer on an amorphous organic layer is easily broken by contact with the probe. The sample was only made directly on a Corning glass substrate, and a 20 nm thick Ag yielded 2.26 $\Omega/\square$.

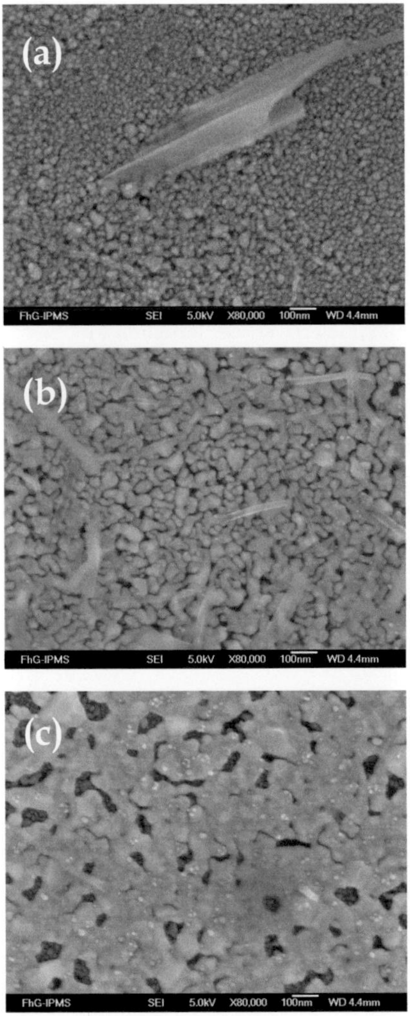

Figure 6.54: SEM images of Ag layer on NET5. The thickness of Ag layer is (a) 20 nm, (b) 30 nm and (c) 40 nm. The SEM images were reprinted from Ref. [186].

The SEM images of the Ag layer on NET5 are shown in Figure 6.54. Interestingly, the morphology of Ag layer in Figure 6.54 (a) showed a completely different layer growth. In particular, the grain size was overall uneven, and some very large grains were sometimes observed. Furthermore, a long-stretched grain was also observed in the middle of the image. In Figure 6.54 (b), such laterally grown structures were found again on a thicker Ag layer. They are presumed as a kind of byproduct from the reaction of Ag and organic layer. The exact mechanism for the reaction is not known, but higher metal-organic complexation of Ag layer than Al by an external photonic excitation or a self-doping effect was demonstrated from the spectra of LDI-TOF-MS observed from Ag on BPhen and Al on BPhen structures [159]. As the thickness of the Ag layer was increased, the crystalline structures seemed to be buried under the Ag layer, as shown in Figure 6.54 (c). However, the Ag layer did not close the organic layer completely even at the thickness of 40 nm. The empty spaces are expected to result in an exposure of organic layers to ambient and render it insufficient as a buffer layer against sputter damage.

Therefore, in order to reduce the reaction of Ag with the organic layer, an extremely thin (~3 nm) Al interlayer was introduced by thermal evaporation before the 15nm thick Ag deposition. Two organic materials NET5 and NET18 were used for the observation of the surface of the Ag layer. These materials are used as electron transport layer. The thickness of the organic layers was 100 nm. In Figure 6.55 (a), the surface of the Ag layer on NET18 is shown. A similar growth of the Ag layer was also observed as on NET5. However, when the thin Al layer was employed between the organic layer and the Ag layer, such a reaction was dramatically reduced as shown in Figure 6.55 (b) and (c). A homogeneous and continuous growth of the Ag layer could be obtained with this interlayer concept, and is attributed to lower reactivity of Al with the organic materials. From the investigation of the Ag morphology, better properties as an electrode and a buffer layer are expected. This thin metal cathode structure was used equally for OLED fabrication before ITO deposition.

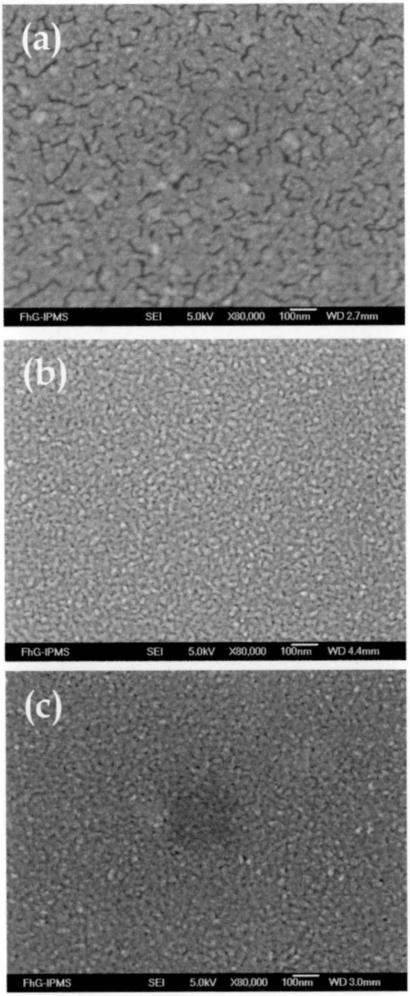

Figure 6.55: SEM images of Ag layers. SEM image of 15 nm thick Ag layer on NET18 is shown in figure (a). Figure (b) and (c) are the images of 15 nm thick Ag layers implementing a 3 nm Al interlayer between the organic layers, NET5 and NET18, and the Ag layers.

6.2.4 Results of OLED with ITO top electrode

Based on the stack listed in Table 5.3, top emission OLEDs were fabricated on the substrate having an Ag anode and silk-printed passivation. In order to evaluate the OLEDs, a reference OLED must be defined first. As examined in section 6.2.1.1, the Ag surface was very vulnerable to oxygen plasma and neutral radicals. Hence only Ar gas was used to treat the Ag surface in the plasma treatment chamber of the cluster equipment. Afterwards, the organic layers were deposited in the organic chambers, and a thin Al/Ag cathode was subsequently deposited in the metal chamber. The OLEDs were then encapsulated by a glass cap with getter. The emission of the reference OLED is shown in Figure 6.56.

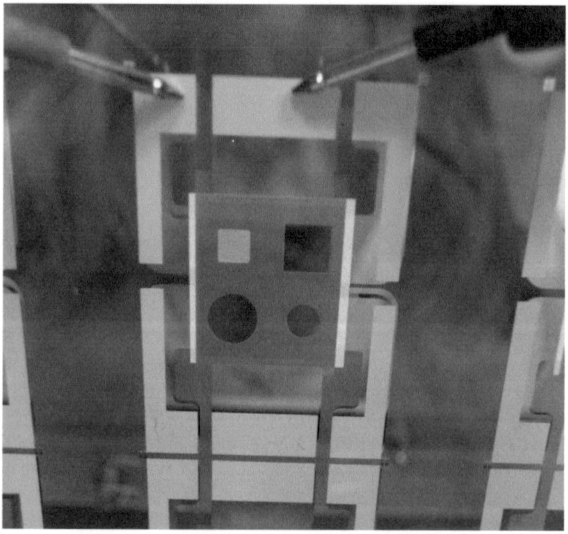

Figure 6.56: Light emission from the reference OLED. The OLED was successfully fabricated and exhibited a homogeneous lighting surface.

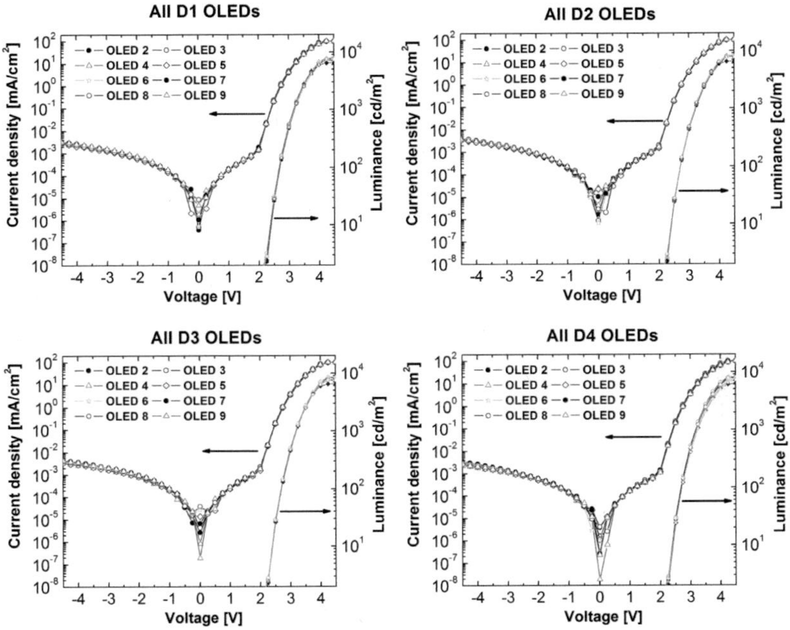

Figure 6.57: *I-V-L* characteristics of the reference top-emission OLEDs. The results of the OLEDs were collected for the size of the OLED from D1 to D4 and shown in separate graphs.

In Figure 6.57, the *I-V-L* characteristics of the reference top-emission OLEDs are exhibited. The fabricated reference OLEDs showed perfectly homogeneous characteristics on the entire substrate. Compared with the leakage currents of p-i-n orange or green OLEDs in section 6.1, slightly higher leakage currents are observed in the present OLEDs. This is however not ascribed to degradation, but a sharper edge of the printed passivation or reduced shunt resistance by the thinner emission layer. The performance of the reference OLED is addressed first.

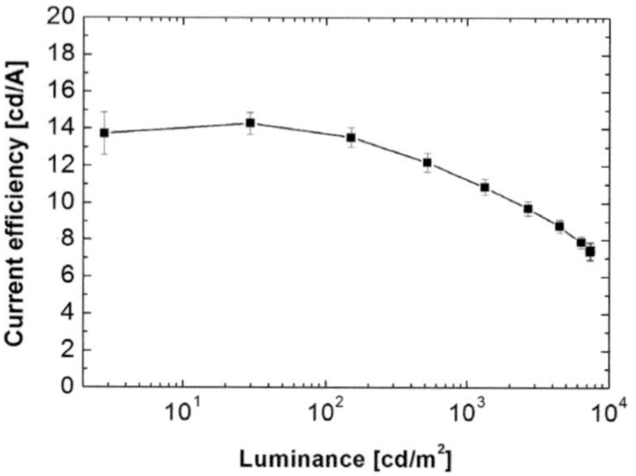

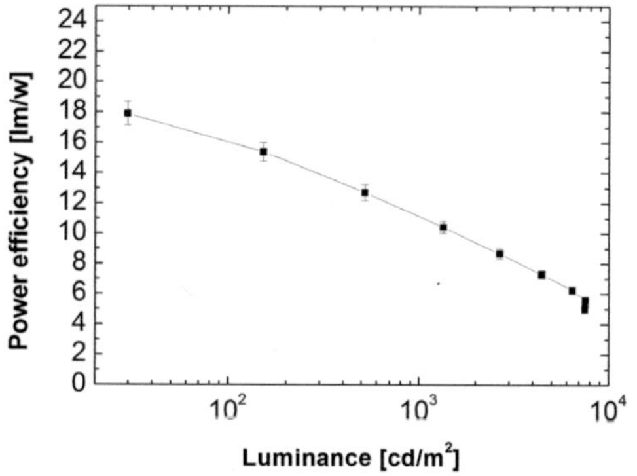

Figure 6.58: Current and power efficiencies of the reference OLEDs with respect to the luminance.

The driving voltage and the efficiencies are listed in Table 6.22. The reference OLED yielded a low driving voltage and moderate efficiencies despite the thin emission layer and the thick transport layers by the help of the optimization of the layer thickness for a good optical outcoupling from the cavity.

Table 6.22. Driving voltage and efficiencies of the fabricated reference OLED

Cathode	at 100 cd/m^2			at 1,000 cd/m^2		
	V [V]	CE [cd/A]	PE [lm/W]	V [V]	CE [cd/A]	PE [lm/W]
Al/Ag	2.67	13.76	16	3.17	11.25	11.11

6.2.4.1 Sputter power mixing ratio variation

In this section, the results for top-emission OLEDs having Al/Ag and ITO top electrode are discussed. 100 nm thick ITO top electrodes were deposited using the process parameters of Table 6.20. In Figure 6.59, *I-V-L* characteristics of the OLEDs out of each substrate are shown. Unfortunately, higher leakage currents and lower luminance were observed from the OLEDs having an ITO top electrode.

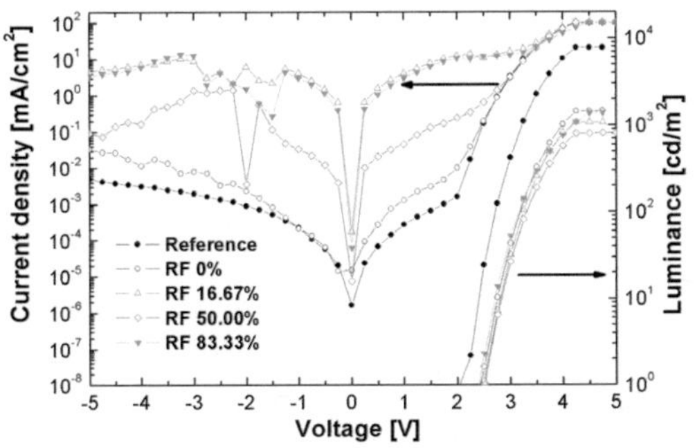

Figure 6.59: *I-V-L* characteristics of representative OLEDs. The OLEDs having an ITO top electrode showed higher leakage current and lower luminance than the reference OLED.

The leakage current values at -5 V were collected and shown in Figure 6.60. While the reference OLEDs showed low and uniform leakage currents, the OLEDs having ITO top electrodes showed much higher and widely distributed leakage currents. Some OLEDs were seriously damaged from the ITO sputtering in spite of the presence of the thin metal electrode, and their leakage currents reached 100 mA/cm^2, which is the maximum compliance of current density. In particular, the leakage currents at the RF portion of 16.67 and 83.33 % appeared higher than at the others. In comparison to the result of the Al sputtering, the overall tendency in Figure 6.60 looked similar to Figure 6.25 except for the RF portion of 16.67 %, where the leakage currents are the highest.

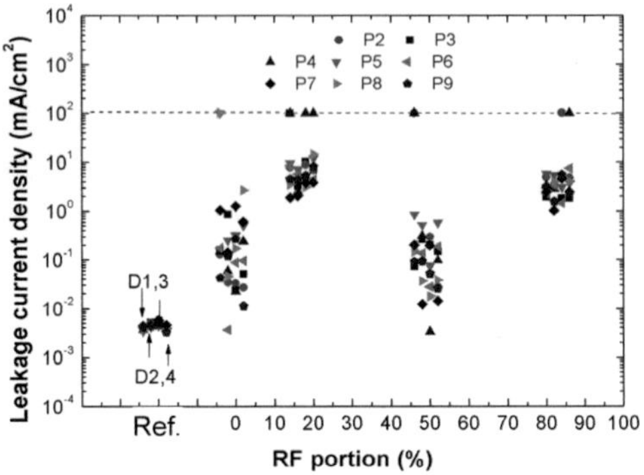

Figure 6.60: Collected leakage current values at -5 V from the OLEDs fabricated.

Simple statistics were introduced for a numerical evaluation. They were carried out with and without the shortcut current densities reaching 100 mA/cm^2, because the shortcut currents increase the mean value considerably. As indicated in Table 6.23, the mean values of the leakage currents including shortcuts showed no remarkable difference in the process parameter variation, but slightly higher values at RF 16.67 %. However, the mean values excluding the shortcuts showed a larger difference, and the best RF portion for low leakage current and shortcut was found at 50 %. The difference between 0 and 50 % was however not so significant except for RF 16.67 %, and these RF portions are the conditions for the DC plasma excitation mode. The damage seemed to decrease as the discharge voltage decreased, but at high RF portion over 50 %, namely in the AC excitation mode, the damage increased abruptly due to an intensive ion bombardment from bulk plasma. Therefore, ITO deposition in the DC excitation mode is preferred for low damage.

Table 6.23. Statistics on the leakage currents. (Unit : mA/cm^2)

Process parameter	Percentage of shortcut (%)	With shortcut		Without shortcut	
		Mean	STD*	Mean	STD*
Reference	0	4.3×10^{-3}	6.11×10^{-3}	4.3×10^{-3}	6.11×10^{-3}
RF 0 %	6.25	6.55	24.1	0.32	0.54
RF 16.67 %	12.5	17.6	31.3	5.89	3.11
RF 50 %	6.25	6.41	24.2	0.17	0.19
RF 83.33 %	6.25	9.5	23.4	3.47	1.49

*STD : standard deviation

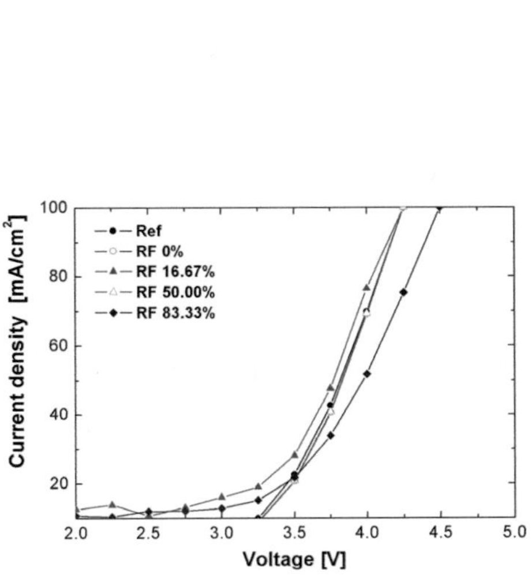

Figure 6.61: *I-V* characteristics of the OLEDs having ITO top electrodes deposited at various RF portions.

Forward *I-V* characteristics were examined as well. The forward current densities are shown in Figure 6.61 in linear scale. Interesting is that the forward current density with the ITO electrodes showed no significant improvement until a RF portion 50 % in comparison to that of the reference OLED. This means that the thin Ag cathode prevented the underlying organic layers from radiation-induced interface damage, but the ITO top electrodes were not helpful to improve the conductivity of the total cathode structure. It is however a natural result if the resistivity of the ITO layers obtained in the former section is considered. The Ag/ITO electrode forms a parallel structure to deliver charge carriers. Since the ITO films on glass processed in the range between 0 and 50 % showed much higher resistivity than the thin Ag electrode, most of the current might flow through the Ag metal cathode. The decrease of the current density at the RF portion 83.33 % is attributed to the degradation of the Ag cathode due to the ion bombardment effect in the AC plasma excitation mode. In order to improve the role of the ITO electrode, a sufficiently low resistance is required. However, it is not easy to achieve the requirement with room temperature-deposited ITO films and thin (~100 nm) thickness in spite of the high RF portion.

In Figure 6.62, the current and power efficiencies are shown with the process parameters, and the OLED performances at 500 cd/m^2 are listed in Table 6.24.

Table 6.24. Driving voltage and efficiencies of the OLEDs at 500 cd/m^2

Process parameter	V [V]	CE [cd/A]	PE [lm/W]
Reference	3.04	12.34	12.77
RF 0 %	3.67	1.55	1.32
RF 16.67 %	3.75	1.04	0.88
RF 50 %	3.75	0.84	0.68
RF 83.33 %	3.93	1.49	1.23

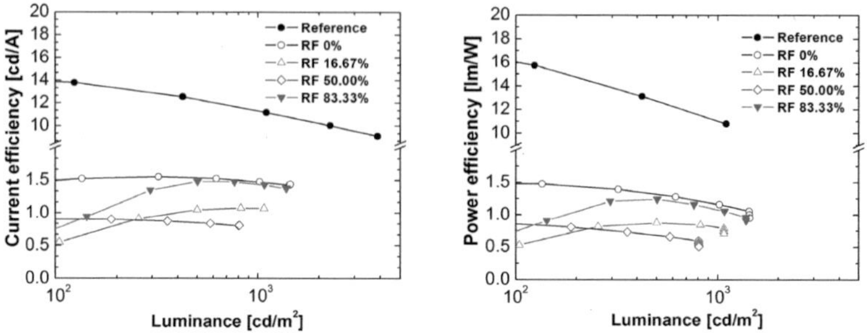

Figure 6.62: Current and power efficiencies of the OLEDs.

The efficiencies were decreased dramatically with the ITO top electrodes. As the RF portion increased, the efficiencies decreased gradually and increased again after the RF portion 50 %. This increase is ascribed to the improved transparency of the ITO layer, but the efficiency at the RF portion 83.33 % was still lower than at the RF 0% despite of the better transparency. This may be due to the decrease of the forward current and the resulting poorer charge balance in the emission layer, which was caused by the deterioration of the Ag cathode. However, such significant decreases of the luminance and efficiency were not expected, and the results can not be explained with only the optical properties of the ITO layers. The minimum transmittance of the ITO film was around 40 % at the wavelength of the photoluminescence maximum of $Ir(ppy)_3$. This means that there might be another reason for the dramatic decrease of the efficiency.

In order to investigate emission characteristics based on thin film optics, the OLED structures were evaluated by means of the ETFOS simulator. The optical constants of the ITO films shown in Figure 6.47 were used for the evaluation, and the results are shown in Figure 6.63. The spectra were obtained by taking only the optical response in the films from the dipole radiation of the emission layer into account. In the inset, integrated radiances are also indicated. The emission characteristics where the RF portions for the ITO deposition were between 0 and 50 % showed approximately half value of that of the reference OLED. However, the OLED with the ITO film deposited at the RF portion of 83.33 % showed slightly higher radiance, and it is attributed to a higher transmittance than other ITO films and resulting better outcoupling effect.

As a method to evaluate the optical response, specular reflectance was measured from the OLEDs using spectrophotometer (Solidspec 3700, Shimadzu scientific instruments USA) at the incident angle of 2.5° from the normal to the surface. In this measurement, the incident light experiences reflection and absorption in the films and the surfaces of the Ag electrodes, and escapes the OLED again. Therefore, the absolute optical losses in the OLEDs are obtained so that their influence to the OLED efficiency can be compared. In Figure 6.64, the measured reflectance curves are shown. The reflectance at the wavelength of 538 nm, where the emission peak lies, was 22.8 % for the reference OLED, and was 16.4, 13.1, 18.2, and 26.2 for the OLEDs with ITO deposited at RF 0, 16.67, 50, and 83.33 %, respectively. The emission simulated and the reflectance measured showed a quite similar tendency in the decrease and increase, and this result suggests that the dramatic decrease of the efficiencies of the OLEDs with the ITO top electrode was caused not only by the optical losses, but probably another effect associated with an electrical loss. Unfortunately, the reason responsible for the low efficiency can not be clarified in this work, and further investigations are necessary.

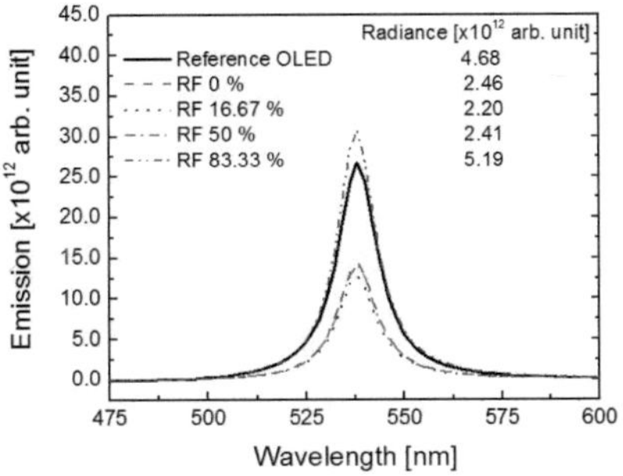

Figure 6.63: Emission characteristic of the OLEDs. Radiance was calculated by integrating the emission spectrum.

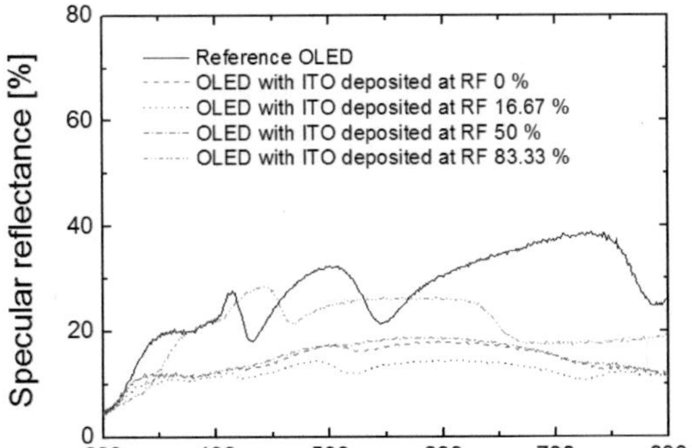

Figure 6.64: Specular reflectance curves measured from the OLEDs. Light fell onto the OLED surface at the incident angle of 2.5° through the encapsulation glass.

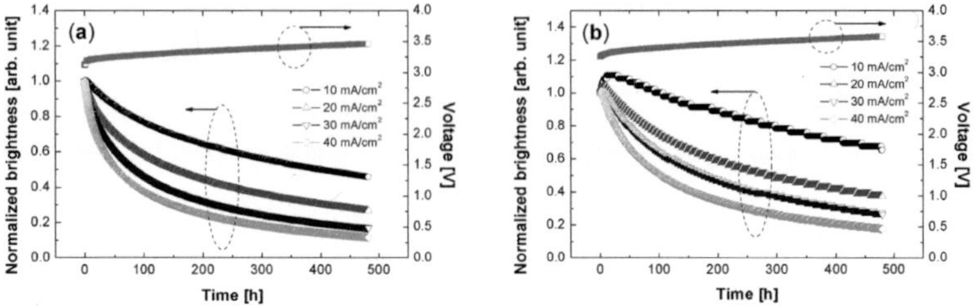

Figure 6.65: Lifetime results of the reference OLED (a) and the OLED having ITO deposited at the RF portion of 0 % (b).

The results of lifetime measurements are shown in Figure 6.65. While the reference OLEDs showed normal decays in initial brightness, the OLEDs with ITO top electrode exhibited an increase in the brightness at the beginning of the measurements. As a result, the lifetime was prolonged, while the driving voltages increased almost comparably. It means that the optical properties of the ITO top electrode might be improved during the measurement through an annealing by the heat generated in the OLED or a local curing by trapped charges in the ITO film. In general, such improvement is known to be possible by applying a sufficiently high external heating and a resulting crystallization. However, it has been also reported that a crystallization of an amorphous ITO film was accomplished by a treatment with a direct electrical current [187]. In the study, a quite high temperature (~200 °C) was induced by the supplement of a small power (1~2 W) to the small sample area (~2 cm^2). The power density on the ITO sample is then five to ten times larger than that of p-i-n type OLED such that a recrystallization of ITO electrode on OLED might occur much slower or less effectively. Since the ITO films seemed to have many defects inside the film, the moving charge carriers on the Ag surface were likely to be trapped in the defects of the ITO film. These trapped charges can induce a localized high electric field or local Joule heating so that a local crystallization might occur. The heat energy might, however, contribute much less because the temperature increase in small OLEDs during operation is usually not as high as over hundred degree.

The evaluated lifetimes are listed in Table 6.25. The lifetime of the OLED with the ITO top electrode was longer than that of the reference OLED, but the sudden death problem still occurred. In particular, the OLEDs having an ITO top electrode deposited at the RF portion of 83.33 % were almost short-circuited alike the Al deposition using FTS, and this result is again attributed to the high leakage current caused by bombardment effect in the AC plasma excitation mode. The two lifetime results indicated as NE* at the RF portion of 16.67 % denote 'not evaluable'. It is because the OLEDs showed unstable operations during the measurement like fluctuation and short circuit. Among the OLEDs successfully evaluated, RF portion 0 and 50 % yielded longer lifetimes than those of the reference OLEDs. If one takes into account the efficiency of OLEDs, it can be estimated that the OLED with ITO deposited at RF 0 % results in a better extrapolated lifetime.

Table 6.25. Measured lifetimes of the fabricated OLEDs

Process parameter	Average lifetime (hour) at current density (mA/cm²)				Sudden death (%) at current density (mA/cm²)			
	10	20	30	40	10	20	30	40
Reference	391.08	156.78	88.18	55.73	0	0	0	0
RF 0 %	651.8	246.98	166.53	91.57	0	0	25	25
RF 16.67 %	NE*	NE*	240.17	174.83	NE*	NE*	25	25
RF 50 %	575.23	303.93	173.45	104.63	0	25	50	0
RF 83.33 %	shortcut	shortcut	shortcut	85.87	100	100	100	50

NE* : not evaluable

The results of extrapolated lifetime on luminance are listed in Table 6.26. The lifetime on the luminance of the RF portion of 0 % showed the longest lifetime. However, the extrapolated lifetimes of the OLEDs having an ITO top electrode were extremely shorter than those of the reference OLEDs, as expected due to the low efficiency. In Figure 6.66, graphs of the lifetime results of the two processes are shown.

Table 6.26. Extrapolated lifetime with respect to luminance

Process parameter	Lifetime (hrs) at 500 cd/m²	Lifetime (hrs) at 1,000 cd/m²
Reference	1,569	500
RF 0 %	125	44
RF 16.67 %	NE*	NE*
RF 50 %	78	35
RF 83.33 %	No data	No data

NE* : not evaluable

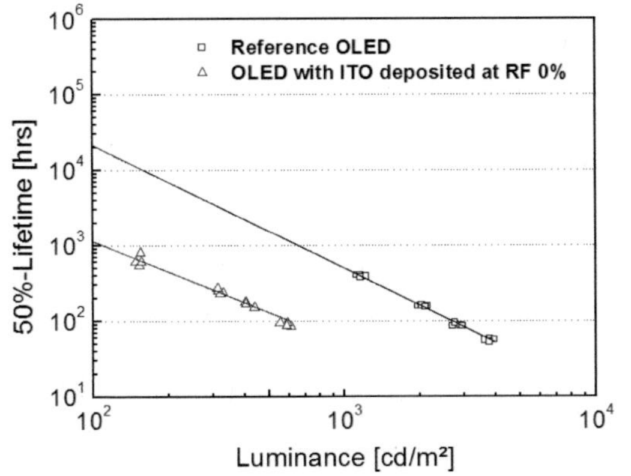

Figure 6.66: Extrapolated lifetime with respect to luminance of the reference OLED and the OLED with ITO top electrode deposited at the RF portion of 0 %.

6.2.4.2 Pressure variation

In this section, the characteristics of top-emission OLED having Al/Ag and ITO top electrode deposited at the process parameters of Table 6.21 are discussed. In Figure 6.67, the *I-V-L* characteristics of four OLEDs taken from each substrate are shown. A higher leakage current and lower luminance were also observed from the OLEDs having an ITO top electrode.

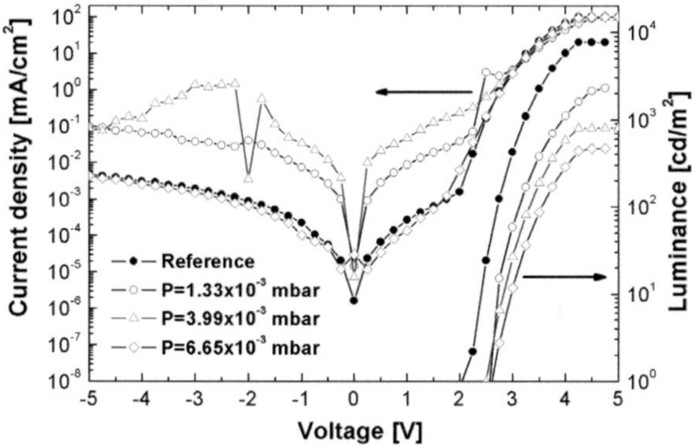

Figure 6.67: *I-V-L* characteristics of the OLEDs fabricated.

Like in previous evaluations, the values of the leakage current were collected and evaluated with simple statistics. In Table 6.27, the percentage of shortcut and the mean values of the leakage currents are listed. By increasing the process pressure, shortcut percentage as well as overall leakage current values decreased gradually. The decrease of the leakage current was obvious but was not so dramatic. This result can be interpreted by an energy loss of particles by collision and a resulting lower energy transfer by the energetic atomic particles.

Table 6.27. Statistics on the leakage currents. (Unit : mA/cm^2)

Process parameter	Percentage of shortcut (%)	With shortcut		Without shortcut	
		Mean	STD*	Mean	STD*
Reference	0	4.3×10^{-3}	6.11×10^{-3}	4.3×10^{-3}	6.11×10^{-3}
1.33×10^{-3}	6.25	6.44	24.2	1.98×10^{-1}	2.79×10^{-1}
3.99×10^{-3}	6.25	6.41	24.2	1.68×10^{-1}	1.88×10^{-1}
6.65×10^{-3}	3.125	3.28	17.4	1.56×10^{-1}	2.18×10^{-1}

*STD : standard deviation

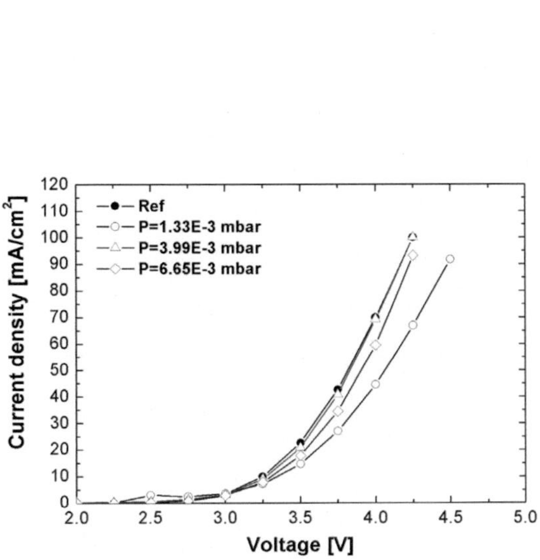

Figure 6.68: *I-V* characteristics of the OLEDs in linear scale, which have ITO top electrode deposited at various process pressures.

The forward *I-V* characteristics were examined as well. In Figure 6.68 the forward current densities are shown in linear scale. A decrease of the forward current density was observed in the OLEDs with ITO deposited at the process pressure of 1.33×10^{-3} mbar. As seen in section 6.2.2, these process parameters resulted in a highly compressive stress in the ITO film, and the reason was explained with less collision of the generated particles possessing a high kinetic energy. The following effect appeared obviously on the surface of the OLED. In Figure 6.69, the emission images of the fabricated OLEDs are exhibited. One can see the formation of buckling waves on the surface of the ITO electrode deposited at the low (1.33×10^{-3} mbar) process pressure. Since no oxygen gas was used, the stress release pattern assures that it was caused mainly by an intrinsic ITO film stress. The compressive stress of the ITO film was approximately 300 MPa. The ITO film would tend to release its stress forming the buckling waves and striping the underlying Ag layer, as shown in Figure 6.38. Therefore, the decrease of the forward current in Figure 6.68 can be attributed to the deterioration of the thin Ag cathode.

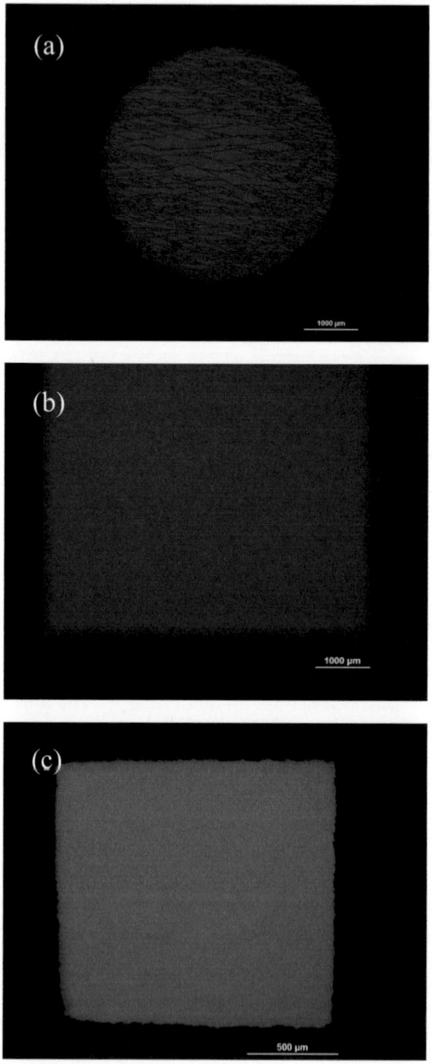

Figure 6.69: Surface images of the OLEDs having ITO top electrode deposited at the process pressure of 1.33×10^{-3} (a), 3.99×10^{-3} (b) and 6.65×10^{-3} (c). In Figure (a), buckling waves were seen on the surface of the ITO film that resulted from the high compressive stress in the film.

The evaluated lifetimes are listed in Table 6.28. Remarkable is that the lifetime of the OLEDs having the ITO top electrode deposited at 1.33×10^{-3} mbar showed the shortest lifetime and highest sudden death rate. This result supports again the degradation of OLED by the high energy of particles. Although the ITO deposition at the high process pressure of 6.65×10^{-3} mbar also resulted in shorter lifetimes, the sudden death ratio was lowest among them. Therefore, high process pressure is helpful to reduce the leakage current and the sudden death problem.

Table 6.28. Measured lifetimes of the fabricated OLEDs

Process parameter	Average lifetime (hour) at current density (mA/cm²)				Sudden death (%) at current density (mA/cm²)			
	10	20	30	40	10	20	30	40
Reference	391.08	156.78	88.18	55.73	0	0	0	0
1.33×10^{-3}	293.27	101.15	38.87	25.2	25	50	50	25
3.99×10^{-3}	575.23	303.93	173.45	104.63	0	25	50	0
6.65×10^{-3}	NE*	109.63	81.73	60.63	NE*	0	25	0

NE* : not evaluable

6.2.4.3 Thickness variation of organic layers

Based on the results from the former two evaluations, an optimal process parameter was established which considers leakage current, efficiency, and lifetime. It used only DC power for better ITO electrical and optical properties at high pressure and longer T-S distance in order to reduce the energy of particles. The process parameter is listed in Table 6.29. The ITO film deposited showed however a high resistivity of 2.15×10^{-2} Ω·cm calculated from its sheet resistance and thickness.

Table 6.29. Process parameters for ITO deposition

RF power (W)	DC power (W)	Ar flow rate (sccm)	Pressure (mbar)	T-T (mm)	T-S (mm)	Deposition time (sec)
0	600	50	6.65×10^{-3}	100	250	5,532

A metal-free OLED was fabricated with only ITO electrode at two different ETL thicknesses. The thickness of the ETL was determined by ETFOS for better outcoupling (see Figure 5.8). 40 and 200 nm thick ETLs were deposited after the HBL (TMM-004), and 100 nm thick ITO electrodes were deposited on them subsequently. In Figure 6.70, the *I-V-L* characteristics of the OLEDs showed extremely high ohmic leakage currents reaching nearly the maximum current compliance and no luminance.

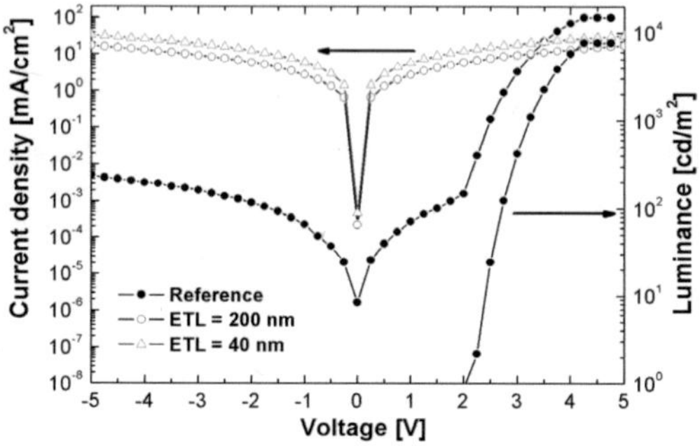

Figure 6.70: *I-V-L* characteristics of the OLEDs fabricated.

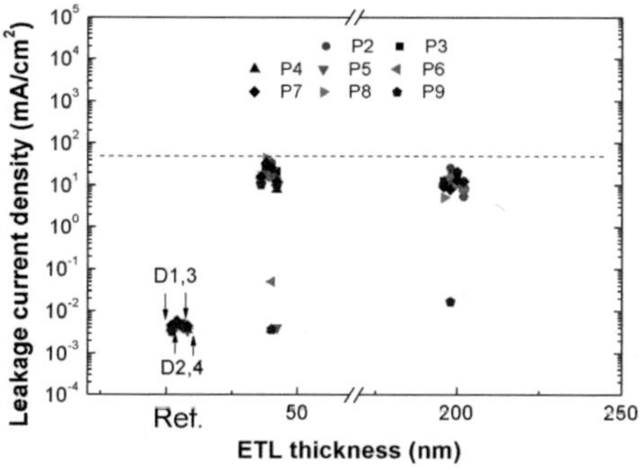

Figure 6.71: Collected leakage current values from the OLEDs.

The leakage current densities were collected and shown in Figure 6.71. The direct ITO deposition on organic layers resulted in a very high leakage current in nearly all OLEDs. Therefore, a protective layer is essentially required before the ITO deposition.

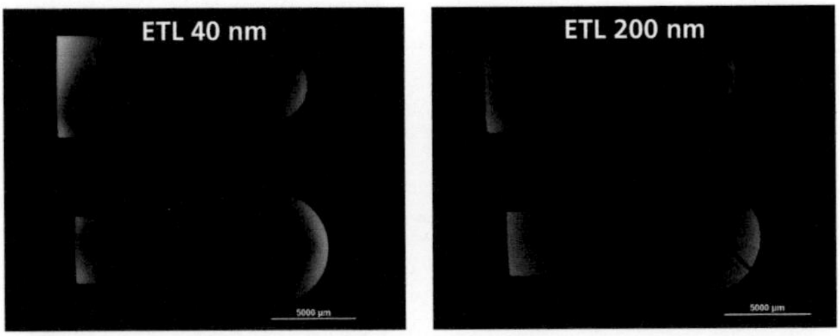

Figure 6.72: Lighting surfaces of the OLEDs where the ITO cathode was directly deposited on organic layers. The figure left is of the OLED with 40 nm thick ETL, and the figure right is with 200 nm thick ETL.

The surface images of operating OLEDs are exhibited in Figure 6.72. Interestingly, partial emission areas in the direction to the cathode contact were observed on the OLED surface. Shortcuts were supposed to exist in the dark areas, and the emissions only at the outer areas were probably due to a low current delivered by a highly resistive ITO electrode. Lifetime measurements were not carried out because no luminance in the middle of the OLED was detected by the photodiode at all, and an extremely short lifetime could be expected.

6.2.4.4 Thickness variation of silver layer

In this section, various thicknesses of the Ag layer were implemented in order to find an optimal Ag layer thickness as a protective layer. The thickness of the Ag layer was varied from 20 to 40 nm. The same process parameter in Table 6.29 was used for 100 nm thick ITO electrode deposition on the 40 nm thick ETL and the Al/Ag cathode. The *I-V-L* characteristics of the OLEDs are exhibited in Figure 6.73.

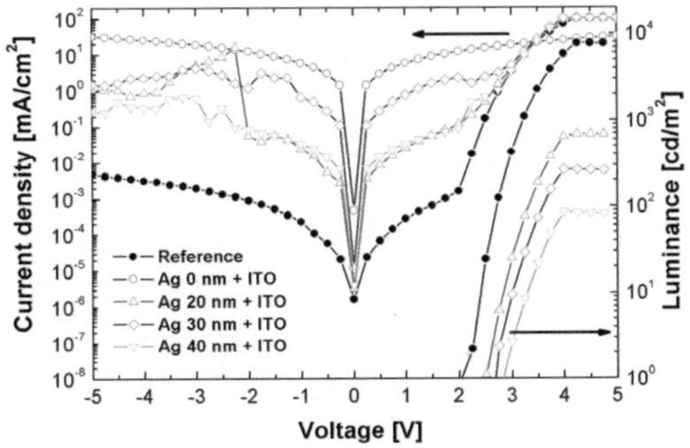

Figure 6.73: *I-V-L* characteristics of the OLEDs that the Ag layer with various thicknesses of 20, 30 and 40 nm was deposited before ITO.

The luminance of the OLEDs decreased strongly as the thickness of the Ag layer increased. This means that the light generated could not leave the cavity structure between the two thick Ag electrodes.

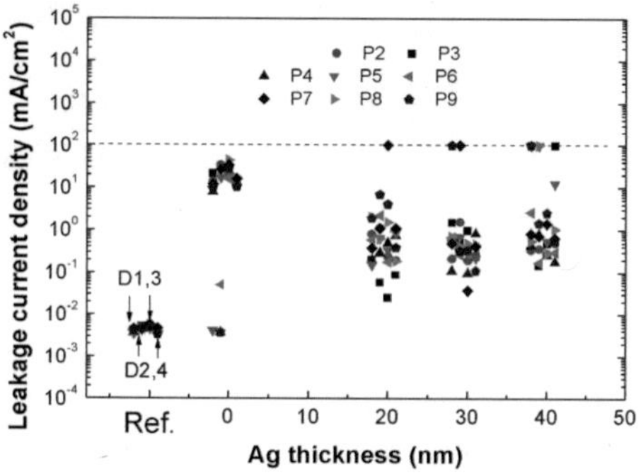

Figure 6.74: Collected leakage current densities from the fabricated OLEDs.

The leakage current densities were collected and are shown in Figure 6.74. Unfortunately, the leakage currents were not decreased with the thicker Ag layer up to 40 nm. Furthermore, the OLEDs with the 40 nm Ag layer exhibited more complete shortcuts after the ITO sputtering. In Table 6.30, the statistics on the leakage currents are listed, and one can see the higher percentage of shortcuts as the thickness of the Ag layer increases. Furthermore, the mean value of the leakage current densities showed also increasing tendency. This result is interpreted that more Ag atoms provided by the thicker layer thickness could contribute to a penetration into the organic layers at locally bombarded places, which can form more number of undesirable short-current paths. Therefore, a thicker metal layer was not so helpful to reduce the leakage current resulted from the impact of the energetic particles.

Table 6.30. Statistics on the leakage current densities. (Unit : mA/cm^2)

Ag thickness (nm)	Percentage of shortcut (%)	With shortcut		Without shortcut	
		Mean	STD	Mean	STD
Reference	0	4.3×10^{-3}	6.11×10^{-3}	4.3×10^{-3}	6.11×10^{-3}
20	3.125	4.07	17.3	0.98	1.35
30	6.25	6.69	24.1	0.47	0.352
40	18.75	19.7	38.6	1.14	2.31

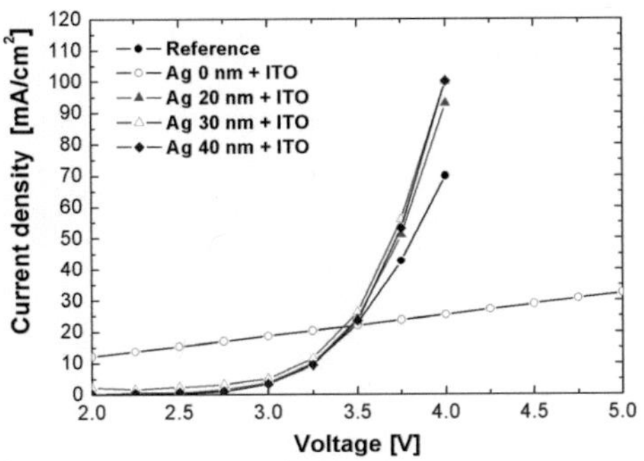

Figure 6.75: *I-V* characteristics of the OLEDs in linear scale with various Ag layer thicknesses. An increase of the current density is observed until the 30 nm thick Ag layer was implemented before the ITO deposition.

Forward *I-V* characteristics of the OLEDs are also shown in Figure 6.75 in linear scale. The forward current density appears to increase further after 3.5 V as the thickness of the Ag layer increases. The result shows that only 15 nm thin top metallic electrode is not sufficiently conductive at high driving voltage above 3.5 V for the top-emission OLED. This is an undesirable limiting factor, in particular, for large area top emission OLED, and is why a highly conductive and transparent top electrode is additionally necessary for large area top-emission and transparent OLEDs. Since the efficiency of the OLEDs was very low, as shown in Figure 6.76, lifetime measurements were also not successful so that the results were not evaluated.

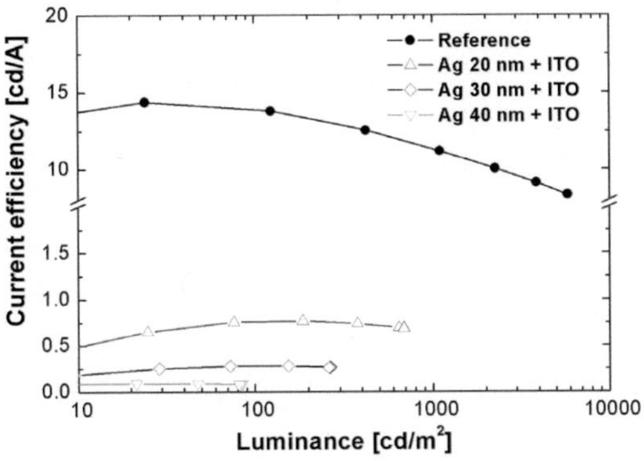

Figure 6.76: Current efficiency of the OLED with various Ag layer thicknesses.

6.2.4.5 Two-step deposition of ITO on OLED

A novel method was attempted in order to satisfy the need for low damage, high conductivity, and transparency of ITO top electrode, based on all results of the former experiments. Conceptual diagrams are given in Figure 6.77.

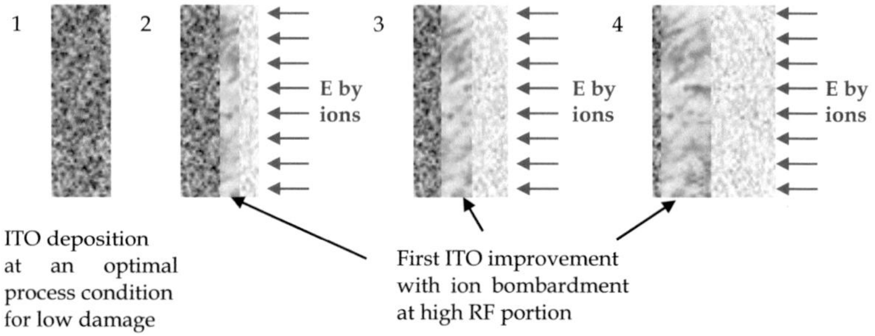

Figure 6.77: Illustration of the concept of the two-step deposition. The first ITO electrode was deposited first with process parameters for low damage (figure 1), and the second ITO electrode is deposited on it with high RF portion, which leads to AC plasma excitation mode and ion bombardment on the substrate.

The first ITO layer was deposited with the process parameters in Table 6.29 for 80 nm thickness. Only DC power was used at high process pressure and longer T-S distance. The second ITO layer was deposited with various thicknesses. RF portion of 83.33 % was used for an effective atomic bombardment, and the process parameters were taken from Table 6.20. The process time was varied for the corresponding thicknesses. As discussed in section 6.2.2, the ion bombardment was beneficial to densify the film, if the ions have moderate energy. Such ion bombardment would modify and improve the properties of the layer, thus a recrystallization or densification of film was anticipated. As the process time of the second ITO deposition became longer, the bombardment effect was expected to appear deeper. The process parameters for the two-step ITO depositions are listed in Table 6.31. The first ITO layer was deposited on the Al/Ag thin cathode at the step number 1. For the second layer, 20, 40, 60, and 80 nm ITO layers were subsequently deposited at the process parameters 2-1, 2-2, 2-3, and 2-4, respectively.

Table 6.31. Process parameters for the two-step ITO deposition

Step No.	RF power (W)	DC power (W)	Ar flow rate (sccm)	Pressure (mbar)	T-T (mm)	T-S (mm)	Deposition time (sec)
1	0	600	50	6.65×10^{-3}	100	250	4,425
2-1	500	100	30	3.99×10^{-3}	100	190	1,019
2-2	500	100	30	3.99×10^{-3}	100	190	2,183
2-3	500	100	30	3.99×10^{-3}	100	190	3,274
2-4	500	100	30	3.99×10^{-3}	100	190	4,366

The modification of the first ITO layer was investigated by means of ellipsometry. Various ITO samples with different second layer thicknesses were made on an Al-coated Corning glass substrate. The samples were measured by the ellipsometer in the standard configuration to obtain the optical constants. An evaluation of depth profiling on the two abrupt ITO layers was carried out at a fixed wavelength of 550 nm. An iterative fitting and a gradient model were used in the WVASE software in order to get a small fitting error.

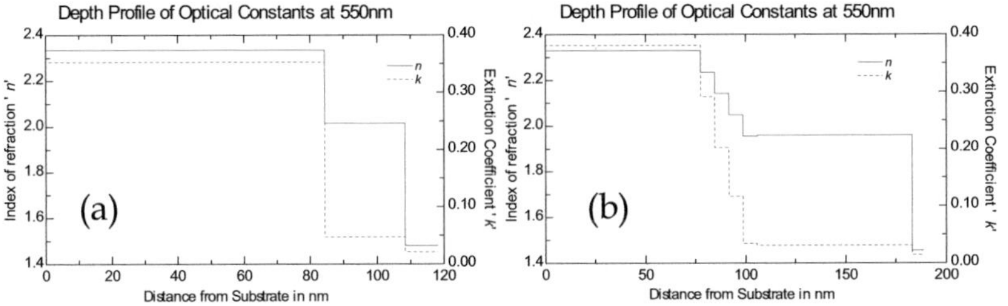

Figure 6.78: Depth profile of optical constants in the abrupt ITO stacks. Figure (a) is from the sample consisting of the first 80 nm ITO and the second 20 nm ITO. Figure (b) is from the first 80 nm ITO and the second 80 nm ITO.

In Figure 6.78, the derived depth profiles of two ITO layers from the step number 2-1 and 2-4 are shown. A clear abrupt step was shown after the second 20 nm ITO layer. It means that the ion bombardment has not influenced the properties of the first ITO layer significantly until the second 20 nm ITO deposition. On the other hand, an interlayer between the two ITO layers was successfully derived after the second 40 nm ITO deposition by adapting a graded model in the fitting work with low mean square error (MSE). These results represents that a modification of the first ITO layer occurred during the second ITO layer and was caused by the bombardment effect from the surface of the first ITO layer. The thicknesses of each interlayer were extracted, and they are listed in Table 6.32 together with the surface roughness of the ITO stacks. The interlayers were derived to be around 30 nm. These total thicknesses measured by ellipsometry were however slightly different from the target thicknesses, and it may be caused by the difference of the measure instruments.

Table 6.32. Thicknesses of the each ITO layer derived by iterative fitting and various models. Fitting errors are also listed.

Step No.	Thickness (nm)			Roughness (nm)	MSE*
	First ITO	Interlayer	Second ITO		
2-1	84.344	0	23.953	9.778	3.662
2-2	67.987	27.195	39.626	14.369	2.915
2-3	75.513	15.568	66.002	2.453	3.874
2-4	70.395	35.199	77.435	5.56	3.695

MSE* : mean square error

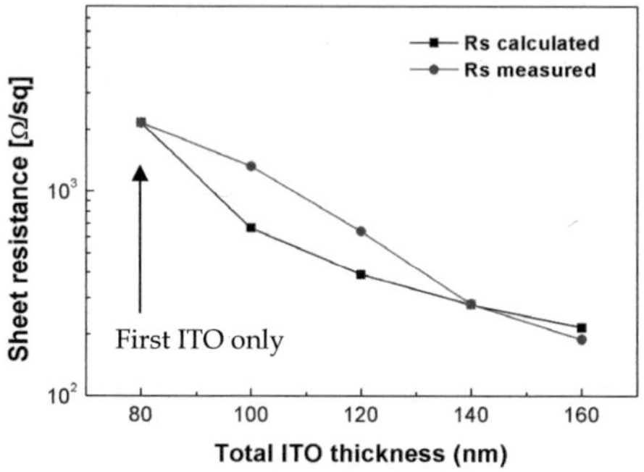

Figure 6.79: Comparison of calculated and measured sheet resistance of ITO layers. A reversal point existed at the total ITO thickness of 140 nm.

The ITO layers deposited on Corning glass substrate were characterized by four point probe in order to evaluate the electrical properties. For an identification of the effect of the interlayers, the sheet resistance was calculated from each resistivity and the thickness of the ITO layers. Since a multilayer consists of parallel resistance components, the sheet resistance was derived from the reciprocal of the summation of the reciprocals of each resistance. In Figure 6.79, the calculated and measured total sheet resistances are shown. A transition point was observed at the total ITO thickness of 140 nm, and the sheet resistance above 140 nm became lower than the calculated one. The differences at the thickness thinner than 140 nm are assigned to an abrupt junction between the two ITO layers. As the irradiation time of the bombardment became longer, the interlayers lowered the total resistance effectively. Eventually, the measured total sheet resistance appeared lower than the calculated one. Therefore, the two-step ITO deposition was expected to be a suitable method for the purposes of this study. Based on the properties of the two-step ITO layers, OLEDs were fabricated with thick ETL (200 nm) and Al/Ag (15nm) thin metal cathode.

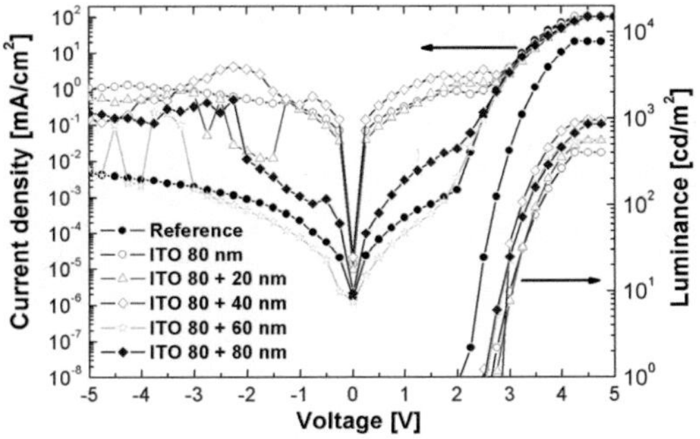

Figure 6.80: *I-V-L* characteristics of the OLEDs having two-step deposited ITO top electrodes.

Figure 6.80 shows the *I-V-L* characteristics of the OLEDs fabricated with two-step deposited ITO top electrodes. Shown are unfortunately still high leakage currents and low luminance. The OLED having the 80 nm thick ITO electrode, which was deposited with the process parameters for a low damage, showed high leakage current density.

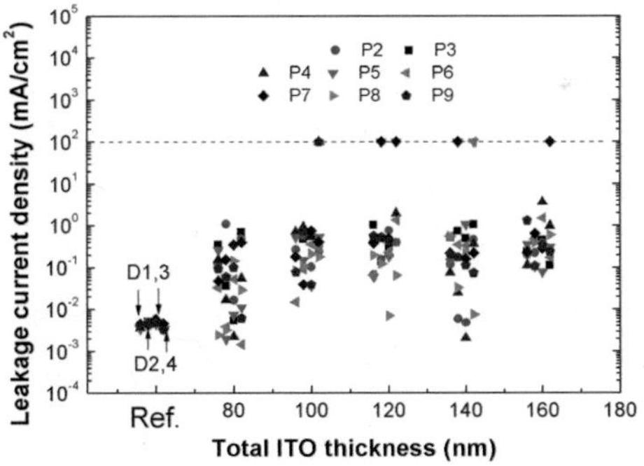

Figure 6.81: Collected leakage current densities at -5 V from the OLEDs with different ITO top electrode thickness from 80 to 160 nm.

The leakage current values at -5 V were collected and plotted in Figure 6.81. The OLEDs with the 80 nm thick ITO electrode also showed widely distributed leakage current densities, but no OLED showed a complete shortcut. One can see the difference by comparing this result with that of Figure 6.60. This is interpreted as the result that the energy of particles was reduced by the higher process pressure and larger T-S distance. However, the OLEDs showed complete shortcuts and higher leakage currents after the second ITO depositions. If one excludes the shortcuts, it is visible that the leakage currents have a upper limit of about 1 mA/cm². Only a slight increase was observed, and this means that the first ITO layer effectively prevented the underlying OLED from the further degradation. One can see the difference of the direct ITO deposition with ion bombardment and the two-step deposition by comparing the leakage current with Figure 6.60.

In Table 6.33 the statistics on the leakage current densities are listed. As mentioned already, the overall leakage currents did not increase considerably despite the ion bombardment on the first ITO surface. The first ITO electrode with different thickness was not tested in this study, but a buffer layer with a sufficiently thick thickness is essentially required in order to protect the OLED.

Table 6.33. Statistics on the leakage current densities. (Unit : mA/cm²)

ITO thickness (nm)	Percentage of shortcut (%)	With shortcut		Without shortcut	
		Mean	STD	Mean	STD
Reference	0	4.3E-3	6.11E-4	4.3E-3	6.11E-4
80	0	1.48E-1	2.33E-1	1.48E-1	2.33E-1
100	6.25	6.55	24.1	3.21E-1	2.58E-1
120	12.5	12.9	32.9	4.07E-1	4.21E-1
140	12.5	12.8	33	2.92E-1	2.87E-1
160	3.125	3.6	17.3	4.9E-1	6.54E-1

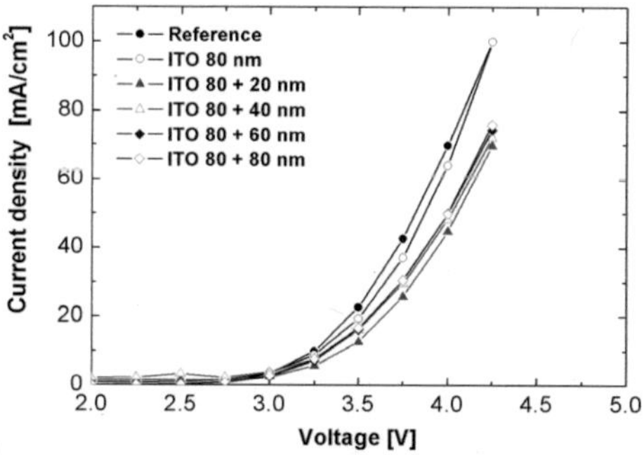

Figure 6.82: Forward *I-V* characteristics of the OLEDs. The second ITO deposition with ion bombardment resulted in a decrease of the forward current.

The forward current densities were also investigated and are shown in Figure 6.82. The forward current density decreased after the second ITO deposition and increased again gradually with increasing the thickness of the total ITO layers. The slow increase of the forward current can be inferred from the improvement of the electrical properties of the ITO layer, but the large decrease in comparison with that of the reference OLED and with other OLED having the single ITO layer is nevertheless not favorable. Although the first ITO layer was helpful to reduce the leakage current in the OLEDs, it probably could not prevent the deterioration of the underlying thin Ag cathode from the atomic bombardment. Unfortunately, the two-step ITO deposition using the ion bombardment was not beneficial to transfer sufficient electrons to the organic layers. A two-step ITO deposition at the RF portion of around 50 % with a well controlled oxygen gas can be a method to improve the property of the total ITO stack instead of the ITO deposition at the high RF portion, but further experiments were not performed.

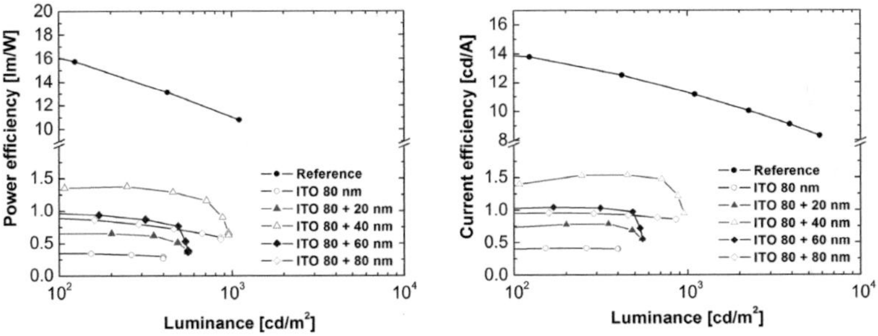

Figure 6.83: Current and power efficiencies of the OLEDs having ITO multilayers with various thicknesses. The maximum efficiency was obtained from the 120 nm thick ITO due to better outcoupling inside the ITO.

Figure 6.83 shows the current and power efficiencies of the OLEDs having ITO multilayers with various thicknesses. The OLEDs showed different efficiencies on the thickness of the ITO stack. The tendency seemed not to be correlated to the variation of the current densities, but rather to be an effect of outcoupling enhancement on capping layer thickness [188]. If the top dielectric capping layer was a stable organic or inorganic materials with good optical properties, much higher efficiency could have been probably obtained. The best efficiency was found at the ITO thickness of 120 nm, and the other OLED parameters are listed in Table 6.34.

Table 6.34. Driving voltage and efficiencies of the fabricated OLEDs

ITO thickness (nm)	at 100 cd/m²			at 400 cd/m²		
	V [V]	CE [cd/A]	PE [lm/W]	V [V]	CE [cd/A]	PE [lm/W]
Reference	2.67	13.76	16	3.17	11.25	11.11
80	3.59	0.39	0.35	4.25	0.41	0.30
100	3.52	0.74	0.66	4.09	0.76	0.58
120	3.24	1.36	1.33	3.69	1.54	1.31
140	3.34	1.04	0.97	3.87	1.01	0.81
160	3.36	0.94	0.89	3.93	0.94	0.74

The lifetime of the OLEDs were evaluated at the current density of 10, 20, 30, and 40 mA/cm^2. Four substrates from each process were used for the measurement at those current densities. The evaluated lifetimes are listed in Table 6.35. The lifetime of the OLEDs with ITO top electrode was longer, but the OLED still showed sudden death problem. Positive was that the OLEDs with a 80 nm thick ITO electrode showed no sudden death problem during the measurement. Although the overall leakage currents observed from the *I-V-L* measurements seemed to be restricted effectively, the ion bombardment effect still resulted in degradation. Therefore it should be noted that the ion bombardment in the AC plasma excitation mode must be avoided.

Table 6.35. Measured lifetimes of the fabricated OLEDs

ITO thickness (nm)	Average lifetime (hour) at current density (mA/cm^2)				Sudden death (%) at current density (mA/cm^2)			
	10	20	30	40	10	20	30	40
Reference	391.08	156.78	88.18	55.73	0	0	0	0
80	407.33	195.13	112.28	85.55	0	0	0	0
100	NE*	NE*	184.95	136.85	NE*	NE*	0	0
120	400.95	205.45	140.43	84.85	50	50	0	0
140	209.6	177.3	134.63	93.65	33.33	33.33	25	0
160	495.83	356.65	254.88	198.5	25	25	0	0

NE*: not evaluable

The extrapolated lifetimes on luminance are listed in Table 6.36. By the help of a better outcoupling effect, the OLEDs having an ITO stack with the thickness of 120 nm yielded a relatively longer lifetime. However, their lifetimes are extremely short, compared with those of the reference OLEDs. The OLEDs with 160 nm thick ITO stack also showed long extrapolated lifetime, because they showed longest lifetimes for half-luminance decay.

Table 6.36. Extrapolated lifetime with respect to luminance

ITO thickness (nm)	Lifetime (hrs) at 500 cd/m^2	at 1,000 cd/m^2
Reference	1,569	500
80	48	32
100	71	35
120	106	49
140	80	48
160	174	122

NE*: not evaluable

6.2.5 Summary and conclusion

Top-emission OLEDs having an ITO top electrode were demonstrated. The ITO top electrodes were deposited by a FTS system in the cluster equipment. As the ITO layer was deposited on organic layers and thin Ag layers, critical deterioration appeared as a form of stress relief pattern like buckling on the surface of the ITO layer. The origin of the deterioration was inferred from a high ITO film stress and an oxidation of the Ag layer, thus various characterization methods were carried out in order to understand the properties of the ITO film on process parameters. RF superimposed DC power was implemented as the power source, and especially oxygen gas was avoided for the ITO deposition in order to prevent the underlying thin Ag layer from oxidation.

By varying the RF portion in the power mixture, it was observed that the electrical and optical properties of the ITO films were strongly correlated. The electrical properties showed a dramatic change, in particular, in carrier mobility between the RF portions of 50 and 83.33 %. The XRD pattern of the ITO film showed a polycrystalline structure at the RF portions from 0 to 50 % and an amorphous structure at 83.33 %, and the crystallinity seemed to follow the variation of discharge voltage. However, it was not easily clarified how an amorphous ITO film was able to yield such a high carrier mobility of 17 $cm^2/V{\cdot}s$ without an extra crystallization procedure. The reason was explained by an additional ion bombardment. As far the Al depositions, a transition of plasma excitation mode was moderately introduced to explain the difference. In the AC plasma excitation mode, a considerable amount of ions bombarding the surface of the substrate is accompanied in FTS system, either they sputter or densify the film. The latter was fit to explain the improved mobility with a film densification by the ion bombardment, thus a very thin depletion layer thickness between crystallites was estimated. A very flat surface was observed from the ITO layer deposited at the high RF portion of 83.33 %, but a highly compressive film stress of 197 MPa was unfortunately found. This compressive stress is a typical example which shows the effect of the ion bombardment.

The process pressure was varied from 1.33×10^{-3} to 6.65×10^{-3} mbar. The electrical properties of the films were, however, not significantly different, but the morphology of the film showed the smoothest surface at the process pressure of 1.33×10^{-3} mbar. This means that the high kinetic energy of the generated particles resulted in an effective densification as well, and the high energy was attributed to less collision with background gases due to low process pressure. Furthermore, a very high compressive stress of almost 300 MPa was found in the film deposited at low process pressure.

Based on the characteristics of the ITO film, top-emission OLEDs having ITO top electrode were fabricated on Ag anode. A reference OLED was made with the combination of Al (3 nm)/Ag (15 nm) in order to prevent the undesirable reaction of Ag with organic layer for a stable thin cathode. As the ITO films were deposited on the OLEDs, serious degradation was unfortunately observed. ITO depositions with varying the RF portion resulted in higher leakage current and lower luminance than those of the reference OLED. Furthermore, the deposited ITO layer was not able to play a role as an electrode on the thin Ag cathode. The ITO deposition in the AC plasma excitation mode at the RF portion of 83.33 % resulted in a decrease of forward current, which means a degradation of the thin Ag cathode by the ion bombardment. Lifetime measurements of the OLEDs were performed, and the results showed a high sudden death ratio. However, the lifetimes of the OLEDs having ITO top electrode were longer than those of the reference OLED. From the measured brightness curves, an increase of the brightness was found at the beginning of the measurement, and this effect yielded a prolonged lifetime for the half decay of brightness. The reason for the improvement of the brightness was ascribed to an annealing effect at the defects of the ITO electrode, especially those at the Ag/ITO interface where the current was delivered. This explanation also gave a clue for the much lower efficiency of the OLEDs with ITO electrodes. The efficiencies of the OLEDs were almost one tenth of those of the reference OLED. Even if the optical properties of the ITO layers were considered, the phenomenon could not be explained with them. Simulated emission characteristics based on thin film optics and reflectance measured in the OLEDs showed only approximately a half decrease compared to those of the reference OLED. There might be, therefore, another reason for the dramatic decrease in the efficiency, and further investigation is necessary.

Various process pressures were also applied to deposit ITO electrodes on the thin Ag cathode and organic layers. Interestingly, ITO deposition at the low process pressure of 1.33×10^{-3} mbar resulted in the buckling degradation on the ITO surface. This means that the high film stress around 300 MPa results in the destruction of the underlying cathode structure by the stress relief. Therefore, the ITO film stress should be controlled to be less than 200 Mpa, if possible.

A direct ITO deposition on organic layers without the Ag cathode was performed with process parameters, which were anticipated to result in a low damage. However, an extremely high leakage current was obtained with partial emission. Thus a sufficiently thick metallic or dielectric buffer layer should be implemented before the ITO deposition. Different thicknesses of the Ag cathode from 20 to 40 nm were also applied in order to find an optimal Ag thickness as a protective layer. However, the OLEDs with the thicker Ag layer exhibited higher leakage currents and shortcut ratio. It is probably because more Ag atoms contributed to a penetration into the organic layers by the bombardment of the energetic particles. Therefore, a metallic layer is not favored to be used as the buffer layer.

Two-step ITO deposition was employed in order to achieve low damage, low resistance, and good transparency. It consisted of a first ITO deposition with the process parameter for a low damage and a second ITO deposition with the high RF power for a highly conductive and transparent film. The second ITO deposition was also expected to improve the properties at the surface of the first ITO layer by the ion bombardment. The formation of an interlayer by the bombardment was evaluated by a depth profiling using ellipsometry. The electrical properties of the two-step ITO deposition showed the effect of the interlayer from the measured sheet resistance. The two-step ITO stacks were successfully applied as the top electrode on the thin Ag cathode. Although the leakage current was not dramatically reduced, the first ITO layer seemed to prevent the OLED from a further degradation by the second ITO deposition. However, the ion bombardment deteriorated the thin Ag cathode and resulted in a decrease of forward current. The efficiency of the OLEDs was at the maximum when a 120 nm thick ITO layer was deposited. An interference effect within the OLED might influence the outcoupling efficiency according to the thickness of the capping ITO layer.

Various experiments have been performed, but the realization of the low-damage ITO deposition on organic layers was not successful so far. However, as a possible method, a dielectric and inorganic protective layer can be applied before the ITO deposition. The requirements for the first electrode are sufficient thickness, high transparency in visible range, high reflection or high absorption in the UV range, low resistivity, and low reactivity with oxygen. Thermally evaporable TCOs are preferred due to their relatively higher conductivity than organic materials. Some undoped TCOs are evaporated at a moderate (600~800°C) temperature in vacuum, such as MoO_3 or In_2O_3 [189-191], so that they are in principle usable in normal OLED equipment. Afterwards, if the ITO electrode is deposited with oxygen in DC plasma excitation mode, good results can be obtained.

7 Concluding remarks and outlook

The main purpose of this study is the development of an alternative method to deposit various top electrodes for large area OLEDs. The magnetron sputtering method was chosen for the purpose, and many experiments were attempted to deposit top electrodes on OLED with low sputter damage. In order to realize low damage sputtering, various power sources and sputter systems were used to deposit metal and TCO material on the top of organic layers, which is necessary as the electrode of BEOLEDs and TEOLEDs, respectively.

In the experiments for the Al metal deposition, the DMS system using bipolar pulsed power showed the best feasibility for the large area deposition as well as low damage sputtering (section 6.1.4), though devices smaller than 1 cm^2 were tested. High deposition rate or short tact time can be achieved by installing several pairs of DMS system in the direction of substrate moving. This method can also be used for organic photovoltaic devices (OPVs) as well as large area OLEDs. Especially, the DMS system will be suitable for highly efficient OLED lighting and OPVs because thick organic stacks consisting of several emission or absorption units, so-called tandem structure, are generally used to improve efficiency. From the viewpoint of hardware, the DMS system is also easy to be applied in mass production equipment, regardless of cluster or inline type. An inline type vacuum chamber with sputter cathodes can be installed between evaporation chambers and encapsulation system so that a continuous OLED fabrication is enabled. Since the process stability and material utilization of magnetron sputter system is superior to that of thermal evaporation, a cost and time-effective production method will be delivered to industry.

In the ITO deposition experiments, two remarkable phenomena were observed: (1) the combination of the ITO film with a thin Ag layer yielded extremely low efficiency in OLEDs, (2) a sufficiently thick pre-deposited inorganic layer seemed to suppress sputter damage from the ITO deposition. Although the reason for the reduction of the efficiency was not clarified by optical measurements, nevertheless it was shown that the thin Ag/ITO combination is not suitable for an effective transparent top electrode. It was also proposed in the summary (section 6.2.5) that an inorganic and transparent conductive material, which is evaporable at a moderate temperature, would be helpful in order to deposit ITO film with low damage in OLED. If TCO film is deposited on organic layers without damage, this would enable various applications like TEOLEDs as well as TOLEDs on glass, metal, or plastic substrates for highly efficient top emission or transparent displays, lightings, etc. TCO top electrodes will be particularly good for white emission, which is composed of at least two colors, due to the weaker microcavity resonance than that of thin metals. However, due to the limits of process temperature, being typically lower than 80°C in OLED fabrication, the TCO films should be deposited at room temperature, and it accompanies a high resistivity of the TCO films. Since the resistance of electrode also affects charge delivery to organic layers in large area OLED, an approach to reduce the resistance is necessary, such as adapting a thin metallic interlayer in between, namely, ITO/Ag or Pd/ITO.

Bibliography

[1] A. Bernanose, M. Comte, and P. Vouaux. "A New Method of Emission of Light by Certain Organic Compounds." J. Chim. Phys. 50, 64 (1953).

[2] C. Tang and S. VanSlyke. "Organic Electroluminescent Diodes." Appl. Phys. Lett. 51, 913 (1987).

[3] R. Meerheim, K. Walzer, M. Pfeiffer, and K. Leo. "Ultrastable and Efficient Red Organic Light Emitting Diodes with Doped Transport Layers." Appl. Phys. Lett. 89, 061111 (2006).

[4] Display search, Austin, Texas. "OLED Lighting to Take Off in 2011." March 12 (2009).

[5] L. S. Liao, L. S. Hung, W. C. Chan, X. M. Ding, T. K. Sham, I. Bello, C. S. Lee, and S. T. Lee. "Ion-beam-induced surface damages on tris-(8-hydroxyquinoline) aluminum." Appl. Phys. Lett. 75, 1619 (1999)

[6] H. Lei, K. Ichikawa, M. Wang, Y. Hoshi, T. Uchida, and Y. Sawada. "Investigation of low-damage sputter-deposition of ITO films on organic emission layer." IEICE Trans. Electron. E91-C, 1658 (2008).

[7] Han-Ki Kim, D.-G. Kim, K.-S. Lee, M.-S. Huh, S. H. Jeong, and K. I. Kim and T. Y. Seong. "Plasma damage-free sputtering of indium tin oxide cathode layers for top-emitting organic light-emitting diodes." Appl. Phys. Lett. 86, 183503 (2005)

[8] Han-Ki Kim, S. W. Kim, K. S. Lee, K. H. Kim. "Direct Al cathode layer sputtering on LiF Alq3 using facing target sputtering with a mixture of Ar and Kr." Appl. Phys. Lett. 88, 083513 (2006)

[9] Yusuke Onai, Takayuki Uchida, Yoshihiro Kasahara, Keisuke Ichikawa, Yoichi Hoshi. "Transparent conductive film for top-emission organic light-emitting devices by low damage facing target sputtering." Thin Solid Films 516, 5911 (2008)

[10] Masahiko Naoe, Shunichi Yamanaka and Youichi Hoshi. "Facing targets type of sputtering method for deposition of magnetic metal films at low temperature and high rate." IEEE Transactions on Magnetics 16, 646 (1980)

[11] M. Hoffman. "Organische Halbleiter." Skript zur Vorlesung, Technische Universität Dresden (2006).

[12] P. Peumans. "Organic thin-film photodiodes." Ph. D thesis, Princeton university (2004).

[13] G. H. Wannier. "The structure of electronic excitation levels in insulating crystals." Phys. Rev. 52, 191 (1937)

[14] J. Frenkel. "On pre-breakdown phenomena in insulators and electronic semi-conductors." Phys. Rev. 54, 647 (1938)

[15] E. A. Silinsh. Organic molecular crystals, their electronic states, Solid-State Science 16, Springer, Heidelberg 1980.

[16] M. Klessinger and J. Michl. Excited states and photochemistry of organic molecules, VCH Publishers, New York (1995).

[17] Rico Meerheim. Quanteneffizienz und Langzeitstabilität monochromer organischer Leuchtdioden. Ph. D dissertation, Technische Universität Dresden (2009).

[18] M. Pope and C. E. Swenberg. Electronic progresses in organic crystals. Oxford university press, Oxford (1982).

[19] M. A. Baldo, D. F. O'Brien, Y. You, A. Shoustikov, S. Sibley, M. E. Thompson, and S. R. Forrest. "Highly efficient phosphorescent emission from organic electroluminescent devices." Nature 395, 151 (1998).

[20] M. A. Baldo, D. F. O'Brien, M. E. Thompson, and S. R. Forrest. "Excitonic singlet-triplet ratio in a semiconducting organic thin film." Phys. Rev. B 60, 14422 (1999).

[21] M. A. Baldo, C. Adachi, and S. R. Forrest. "Transient analysis of organic electrophosphorescence. II. Transient analysis of triplet-triplet annihilation." Phys. Rev. B 62, 10967 (2000).

[22] T. Förster. Fluoreszenz organischer Verbindungen. Vandenhoek und Ruprecht, Göttingen (1951).

[23] M. A. Baldo, M. E. Thompson, S. R. Forrest. "High-efficiency fluorescent organic light-emitting devices using a phosphorescent sensitizer." Nature. 403, 750 (2000).

[24] U. Albrecht and H. Bässler. "Efficiency of charge recombination in organic light emitting diodes." Chem. Phys. 199, 207 (1995).

[25] J. Blochwitz, M. Pfeiffer, T. Fritz, K. Leo, D. M. Alloway, P. A. Lee, N. R. Armstrong. "Interface electronic structure of organic semiconductors with controlled doping levels." Org. Electron. 2, 97 (2001).

[26] S. L. Lai, M. Y. Chan, M. K. Fung, C. S. Lee, L. S. Hung and S. T. Lee. "Applications of ytterbium in organic light-emitting devices as high performance and transparent electrodes." Chem. Phys. Lett. 366, 128 (2002).

[27] C. J. Lee, J. I. Han, D. G. Moon. "Transparent white OLEDs using Ca/Ag cathode." Mol. Cryst. Liq. Cryst. 459, 75 (2006).

[28] H. Bässler. "Charge transport in disordered organic photoconductors: a Monte carlo simulation study." Phys. Stat. Sol B 175, 15 (1993).

[29] J. Staudigel, M. Stößel, F. Steuber, and J. Simmerer. "A quantitative numerical model of multilayer vapor-deposited organic light emitting diodes." J. Appl. Phys. 86, 3895 (1999).

[30] W. Brütting, S. Berleb, and A. G. Mückl. "Space-charge limited conduction with a field and temperature dependent mobility in Alq light-emitting devices." Synth. Met. 122, 99 (2001).

[31] W. Brütting. "Physics of organic semiconductors." John Wiley & Sons (2005).

[32] R. G. Kepler, P. M. Beeson, S. J. Jacobs, R. A. Anderson, M. B. Sinclair, V. S. Valencia, and P. A. Cahill. "Electron and hole mobility in tris(8-hydroxyquinolinolato-N1,O8) aluminum." Appl. Phys. Lett. 66, 3618 (1995).

[33] B. J. Chen, W. Y. Lai, Z. Q. Gao, C. S. Lee, S. T. Lee, and W. A. Gambling. "Electron drift mobility and electroluminescent efficiency of tris(8-hydroxyquinolinolato) aluminum." Appl. Phys. Lett. 75, 4010 (1999).

[34] A. Ioannidis, E. Forsythe, Y. Gao, and M. W. Wu. "Current-voltage characteristic of organic light emitting diodes." Appl. Phys. Lett. 72, 3038 (1998).

[35] P. E. Burrows and S. R. Forrest. "Electroluminescence from trap-limited current transport in vacuum deposited organic light emitting devices." Appl. Phys. Lett. 64, 2285 (1994).

[36] L. S. Hung, C. W. Tang, M. G. Mason, P. Raychaudhuri, J. Madathil. "Application of an ultrathin LiF/Al bilayer in organic surface-emitting diodes." Appl. Phys. Lett. 78, 544 (2001).

[37] R. G. Kepler, P. M. Beeson, S. J. Jacobs, R. A. Anderson, M. B. Sinclair, V. S. Valencia, and P. A. Cahill. "Electron and hole mobility in tris(8-hydroxyquinolinolato-N1, O8) aluminum." Appl. Phys. Lett. 66, 3618 (1995).

[38] P. E. Burrows, Z. Shen, V. Bulovic, D. M. McCarty, and S. R. Forrest. "Relationship between electroluminescence and current transport in organic heterojunction light-emitting devices." J. Appl. Phys. 79, 7991 (1996).

[39] V. I. Adamovich, S. R. Cordero, P. I. Djurovich, A. Tamayo, M. E. Thompson, B. W. D'Andrade, and S. R. Forrest. "New charge-carrier blocking materials for high efficiency OLEDs." Org. Electron. 4, 77 (2003).

[40] K. Walzer, B. Maennig, M. Pfeiffer and K. Leo. "Highly efficient organic devices based on electrically doped transport layers." Chem. Rev. 107, 1233 (2007).

[41] L. S. Hung and C. H. Chen. "Recent progress of molecular organic electroluminescent materials and devices." Mater. Sci. Eng., R 39, 143 (2002).

[42] H. Kim, A. Pique, J. S. Horwitz, H. Murata, Z. H. Kafafi, C. M. Gilmore, D. B. Chrisey. "Effect of aluminum doping on zinc oxide thin films grown by pulsed laser deposition for organic light-emitting devices." Thin Solid Films. 377, 798 (2000).

[43] J. Cui, A. Wang, N. L. Edleman, J. Ni, P. Lee, N. R. Armstrong, T. J. Marks. "Indium Tin Oxide alternatives - High work function transparent conducting oxides as anodes for organic light-emitting diodes." Adv. Mater. 13, 1476 (2001).

[44] J. Kalinowski and K. Szybowska. "Injection of charge into the archetype organic hole transporting material TPD at the electrical contacts with ITO and Al." Org. Electron. 9, 883 (2008).

[45] W. Brütting, s. Berleb, and A. G. Mückl. "Device physics of organic light-emitting diodes based on molecular materials." Org. Electron. 2, 1 (2001).

[46] D. F. O'Brien, M. A. Baldo, M. E. Thompson, S. R. Forrest. "Improved energy transfer in electrophosphorescent devices."Appl. Phys. Lett. 74, 442 (1999).

[47] X. Jiang, Z. Zhang, W. Zhao, W. Zhu, B. Zhang, and S. Xu. "White-emitting organic diode with a doped blocking layer between hole- and electron-transporting layers." J. Phys. D: Appl. Phys. 33, 473 (2000).

[48] J. Kido, C. Ohtaki, K. Hongawa, K. Okuyama, and K. Nagai. "1,2,4-Triazole derivative as an electron transport layer in organic electroluminescent devices." Jpn. J. Appl. Phys. 32, L917 (1993).

[49] G. Pathasarathy, P. E. Burrows, V. Khalfin, V. G. Kozolov, S. R. Forrest. "A metal-free cathode for organic semiconductor devices." Appl. Phys. Lett. 72, 2138 (1998).

[50] L. S. Hung, C. W. Tang. "Interface engineering in preparation of organic surface-emitting diodes." Appl. Phys. Lett. 74, 3209 (1999).

[51] G. Pathasarathy, C. Adachi, P. E. Burrows, S. R. Forrest. "High-efficiency transparent organic light-emitting devices." Appl. Phys. Lett. 76, 2128 (2000).

[52] P. E. Burrows, G. Gu, S. R. Forrest, E. P. Vicenzi, T. X. Zhou. "Semitransparent cathodes for organic light emitting devices." J. Appl. Phys. 87, 3080 (2000).

[53] Z. Meng, C. H. Chen, C. Qiu, H. S. Kwok, and M. Wong. "Active-matrix organic light-emitting diode display implemented using metal-induced unilaterally crystallized polycrystalline silicon thin-film transistors." SID symp. Digest. 32, 380 (2001).

[54] R. M. A. Dawson, Z. Shen, D. A. Furst, S. Connor, J. Hsu, M. G. Kane, R. G. Stewart, A. Ipri, C. S. King, P. J. Green, R. T. Flegal, S. Pearson, W. A. Barrow, E. Dickey, K. Ping, C. W. Tang, S. A. VanSlyke, F. Chen, J. Shi, J. C. Strum, and M. H. Lu. "Design of an improved pixel for a polysilicon active-matrix organic LED display." SID symp. Digest. 29, 11 (1998).

[55] R. M. A. Dawson, Z. Shen, D. A. Furst, S. Connor, J. Hsu, M. G. Kane, R. G. Stewart, A. Ipri, C. S. King, P. J. Green, R. T. Flegal, S. Pearson, W. A. Barrow, E. Dickey, K. Ping, S. Robinson, C. W. Tang, S. A. VanSlyke, F. Chen, J. Shi, M. H. Lu, and J. C. Strum. "Design of an improved pixel for a polysilicon active-matrix organic LED display." IEEE IEDM Technical digest. 875 (1998).

[56] C. J. Lee, R. B. Pode, D. G. Moon, J. I. Han. "Realization of an efficient top emission organic light-emitting device with novel electrodes." Thin Solid Films. 467, 201 (2004).

[57] Q. Huang. Ph. D dissertation, High efficiency top-emitting organic light-emitting diodes: Design and fabrication. Technische Universität Dresden (2007).

[58] A. Dodabalapur, L. J. Rothberg, R. H. Jordan, T. M. Miller, R. E. Slusher, and Julia M. Phillips. "Physics and applications of organic microcavity light emitting diodes." J. Appl. Phys. 80, 6954 (1996).

[59] J. H. Cheon, J. H. Choi, J. H. Hur, J. Jang, H. S. Shin, J. K. Jeong, Y. G. Mo, and H. K. Chung. "Active-matrix OLED on bendable metal foil." IEEE Trans. Electron Devices. 53, 1273 (2006).

[60] Y. Hong, S. Chung, A. Z. Kattamis, I. C. Cheng, and S. Wagner. "Technical issues of stainless steel foil substrates for OLED display applications." SPIE, Proc. SPIE, 6655, 66550N (2007).

[61] J. Meyer, T. Winkler, S. Hamwi, S. Schmale, M. Kröger, P. Görrn, H. H. Johannes, T. Riedl, E. Lang, D. Becker, and T. Kowalsky. "Highly efficient fully transparent inverted OLEDs." SPIE, Proc. SPIE, 6655, 66550N (2007).

[62] U. Hoffmann, P. Netuschill, M. Bender, P. Sauer, M. Schreil, J. Amelung, and K. Leo. "OLED manufacturing using vertical in-line machine concept." SID Symp. Dig. Tech. Papers 34, 1410 (2003).

[63] C. May, Y. Tomita, M. Toerker, M. Eritt, F. Loeffler, J. Amelung, and K. Leo. "In-Line deposition of organic light-emitting devices for large area applications." Thin Solid Films 516, 4609 (2007).

[64] T. Wakimoto, Y. Fukuda, K. Nagayama, A. Yokoi, H. Nakada, and M. Tsuchida. "Organic EL cells using alkaline metal compounds as electron injection materials." IEEE Trans. Electron. Devices 44, 1245 (1997).

[65] L. S. Hung, C. W. Tang, M. G. Mason. "Enhanced electron injection in organic electroluminescence devices using an Al/LiF electrode." Appl. Phys. Lett. 70, 152 (1997).

[66] K. L. Chopra, S. Major and D. K. Pandya. "Transparent conductors-A status review." Thin solid films. 102, 1 (1983)

[67] Y. Tomita, C. May, M. Toerker, J. Amelung, M. Eritt, F. Loeffler, C. Luber, K. Leo, K. Walzer, K. Fehse, and Q. Huang. "Highly efficient p-i-n-type organic light emitting diodes on ZnO:Al substrates." Appl. Phys. Lett. 91, 063510 (2007)

[68] K. Schulze, B. Männig, Y. Tomita, C. May, J. Hüpkes, E. Brier, E. Reinold, P. Bäuerle, and K. Leo. "Organic solar cells on indium tin oxide and aluminium doped zinc oxide anodes." Appl. Phys. Lett. 91, 073521 (2007)

[69] H. P. Löbl, M. Huppertz and D. Mergel. "ITO films for antireflective and antistatic tube coating sprepared by d.c. magnetron sputtering." Surf. Coat. Technol. 82, 90 (1996)

[70] I. Hamberg and C. G. Granqvist. "Evaporated Sn-doped In2O3 films: Basic optical properties and applications to energy-efficient windows." J. Appl. Phys. 60, R123 (1986)

[71] R. L. Weiher. "Electrical properties of single crystals of indium oxide." J. Appl. Phys. 33, 2834 (1962)

[72] N. F. Mott, Metal-Insulator transitions (Taylor and Francis, London, 1974)

[73] P. P. Edwards and M. J. Sienko. "Universality aspects of the metal-nonmetal transition in condensed media." Phys. Rev. B 17, 2575 (1978)

[74] G. D. Mahan. "Energy gap in Si and Ge: Impurity dependence." J. Appl. Phys. 51, 2634 (1980)

[75] K. F. Berggren and B. E. Sernelius. "Band-gap narrowing in heavily doped many-valley semiconductors." Phys. Rev. B 24, 1971 (1981)

[76] E. Burstein. "Anomalous optical absorption limit in InSb." Phys. Rev. 93, 632 (1954)

[77] T. S. Moss. "The interpretation of the properties of indium antimonide." Proc. Phys. Soc. B 67, 775 (1954)

[78] I. Hamberg, C. G. Granqvist, K. F. Berggren, B. E. Sernelius and L. Engström. "Band-gap widening in heavily Sn-doped In2O3." Phys. Rev. B 30, 3240 (1984)

[79] Y. Ohhata, F. Shinoki and S. Yoshida. "Optical properties of r.f. reactive sputtered tin-doped In2O3 films." Thin solid films. 59, 255 (1979)

[80] F. T. J. Smith and S. L. Lyu. "Effects of heat-treatment on indium-tin oxide films." J. electrochem. Soc. 128, 2388 (1981)

[81] F. Urbach. "The long-wavelength edge of photographic sensitivity and of the electronic absorption of solids." Phys. Rev. 92, 1324 (1953)

[82] W. Franz. "Einfluß eines elektrischen Feldes auf eine optische Absorptionskante." Z. Naturforsch. 13a, 484 (1958).

[83] L. V. Keldysh. "Behaviour of non-metallic crystals in strong electric fields." Soviet Physics JETP 6, 763 (1958).

[84] F. A. Kroger. "Chemistry of imperfect crystals." North-Holland, Amsterdam. Cahps. 7, 16 (1964).

[85] C. A. Vincent. "The nature of semiconductivity in polycrystalline tin oxide." J. Electrochem. Soc. 119, 515 (1972).

[86] J. C. C. Fan and J. B. Goodenough. "X-ray photoemission spectroscopy studies of Sn-doped indium-oxide films." J. Appl. Phys. 48, 3524 (1977).

[87] R. G. Gordon. "Criteria for choosing transparent conductors." MRS Bull. 25, 52 (2000).

[88] G. J. Exarhos and X. Zhou. "Discovery-based design of transparent conducting oxide films." Thin Solid Films 515, 7025 (2007).

[89] R. L. Petritz. "Theory of photoconductivity in semiconductor films." Phys. Rev. 104, 1508 (1956).

[90] J. Y. W. Seto. "The electrical properties of polycrystalline silicon films." J. Appl. Phys. 46, 5247 (1975).

[91] D. H. Zhang. "Fast photoresponse and the related change of crystallite barriers for ZnO films deposited by RF sputtering." J. Phys. D 28, 1273 (1995).

[92] E. Shanthi, V. Dutta, A. Banerjee and K. L. Chopra. "Electrical and optical properties of undoped and antimony-doped tin oxide film." J. Appl. Phys. 51, 6243 (1980)

[93] E. Shanthi, A. Banerjee, V. Dutta, and K. L. Chopra. "Annealing characteristics of tin oxide films prepared by spray pyrolysis." Thin Solid Films 71, 237 (1980).

[94] C. R. Dutta and K. Barua. "Space-charge limited conduction in praseodymium oxide films." J. Phys. D: Appl. Phys. 14, L179 (1981).

[95] C. Adachi, M. A. Baldo, M. E. Thompson, and S. R. Forrest. "Nearly 100% internal phosphorescence efficiency in an organic light emitting device," J. Appl. Phys. 90, 5048 (2001).

[96] M. Knudsen. "The molecular streaming and effusion of gases through openings." Ann. Phys. Lpz. 28, 999 (1909).

[97] P. E. Burrows, S. R. Forrest, L. S. Sapochak, J. Schwartz, P. Fenter, T. Buma, V. S. Ban, J. L. Forrest. "Organic vapor phase deposition: a new method for the growth of organic thin films with large optical non-linearities." J. Cryst. Growth 156, 91 (1995).

[98] M. Eritt. "Großflächige Abscheidung organischer Leuchtdioden und Nutzung optischer Verfahren zur in-situ Prozesskontrolle.", in publication, TU Dresden (2010)

[99] K. Fehse. "Organische Leuchtdioden mit Polymeranoden." PhD thesis, TU Dresden (2007).

[100] Y. Tomita. "Alternative transparent electrodes for organic light emitting diodes." PhD thesis, TU Dresden (2008).

[101] A. J. Heeger, W. R. Salaneck, I. Lundstrom and B. Ranby. "Conjugated polymers and related materials." Oxford, 27-62 (1993).

[102] S. L. Lai, M. Y. Chan, M. K. Fung, C. S. Lee, L. S. Hung and S. T. Lee. "Applications of Ytterbium in organic light-emitting devices as high performance and transparent electrodes." Chem. Phys. Lett. 366, 128 (2002).

[103] F. Ventsch, M. C. Gather and K. Meerholz. "Towards organic light-emitting diode microdisplays with sub-pixel patterning." Org. Electron. 11, 57 (2010)

[104] C. H. Tsai, M. S. Chang, C. C. Liu, S. Y. Chu, C. I. Hung and M. P. Houng. "Enhancement of the luminance efficiency for top-emitting DB-PPV polymer light-emitting diodes with Ca/Ag cathode." Mater. Chem. Phys. 111. 158 (2008).

[105] V. I. Kostikov, Yu. I. Koshelev, E. F. Filimonov, E. M. Tatievskaya and R. N. Ponkratova. "Wetting of carbeon-graphite materials by liquid aluminum." Translated from Poroshkovaya Metallurgiya, 10, 79 (1981)

[106] S. Mayo, K. F. Galloway and T. F. Leedy. "Radiation dose due to electron-gun metallization systems." IEEE Trans. Nucl. Sci. 23, 1875 (1976).

[107] H. S. Pang, Y. S. Jeong, H. S. Choi, C. W. Han, Y. H. Tak and B. C. Ahn. "Effects of x-ray radiation on hydrogenated amorphous silicon thin-film transistors and characteristics of tungsten shield layer." Curr. Appl. Phys. 10, 184 (2010).

[108] U. Hoffmann, P. Netuschill, P. Sauer, H. G. Lotz, A. Klöppel, M. Bender, J. Trube. "New In-line machine concept for OLRF manufacturing." SID Symp. Dig. Tech. Papers 33, 891 (2002).

[109] M. A. Lieberman. Principles of plasma discharges and materials processing. John Wiley & Sons, 1994.

[110] B. Chapman, Glow discharge process, John Wiley & Sons, 1980.

[111] Atomic collision group in the department of physics, Wisconsin university, USA. http://raptor.physics.wisc.edu.

[112] J. W. Carr, M. W. Blades 'Emission spectra of an argon inductively coupled plasma in the vacuum ultraviolet: background spectra from 85 to 200 nm' Spectrochimica Acta Part B: Atomic Spectroscopy, 39, 567-574 (1984)

[113] J. L. Zhang, X. L. Deng, P. S. Wang, T. C. Ma, 'Emission spectrum diagnostics of argon DC discharge' Vacuum 59, 80 (2000)

[114] M. Ohring. "The materials science of thin films." Academic press, Inc. (1992)

[115] J. A. Thornton. "Influence of apparatus geometry and deposition conditions on the structure and topography of thick sputtered coatings." J. Vac. Sci. Technol. 11, 666 (1974).

[116] D. M. Mattox. Handbook of physical vapor deposition (PVD) processing. Noyes Publications, 1998.

[117] S. M. Rossnagel. "Energetic particle bombardment of films during magnetron sputtering." J. Vac. Sci. Technol A7, 1025 (1989).

[118] J. C. Helmer, C. E. Wickerscham. "Pressure effects in planar magnetron sputter deposition." J. Vac. Sci. Technol. A 4 (3), 408 (1986).

[119] Tilo P. Drüsedau, Mirko Löhmann, and Bernd Garke. "Decay length of the pressure dependent deposition rate for magnetron sputtering." J. Vac. Sci. Technol. A 16 (4), 2728 (1998).

[120] A. Settaouti, L. Settaouti. "Simulation of the transport of sputtered atoms and effects of processing conditions." Appl. Surf. Sci. 254, 5750 (2008).

[121] S. D. Ekpe, and S. K. Dew, 'Investigation of thermal flux to the substrate during sputter deposition of aluminum' J. Vac. Sci. Technol. A 20, 1877 (2002).

[122] K. Ishibashi, K. Hirata and N. Hosokawa. "Mass spectrometric ion analysis in the sputtering of oxide targets." J. Vac. Sci.Technol. A10, 1718 (1992).

[123] K. Tominaga, T. Yasuda, M. Kume and O. Tada. "Influence of energetic oxygen bombardment on conductive ZnO films." Jpn. J. Appl. Phys. 24, 944 (1985).

[124] Y. Hoshi and R. Ohki. "Low energy rf sputtering system for the deposition of ITO thin films." Electrochimica Acta 44, 3927 (1999).

[125] K. Ellmer, R. Cebulla, and R. Wendt. "Transparent and conducting ZnO:Al films deposite by simultaneous RF- and DC-excitation of a magnetron," Thin Solid Films 317, 413 (1998)

[126] K. Ellmer, R. Cebulla, and R. Wendt, "Characterization of a magnetron sputtering discharge with simultaneous RF- and DC-excitation of the plasma for deposition

of transparent and conductive ZnO:Al films," Surface & Coatings Technology 98, 1251 (1998)

[127] M. Bender, J. Trube and J. Stollenwerk, "Deposition of transparent and conducting indium-tin-oxide films by the RF superimposed DC sputtering technology," Thin Solid Films 354, 100 (1999).

[128] Advanced Energy. "Arc handling in RF superimposed DC processes application note." Application notes (2007).

[129] R. Scholl, A. Belkind and Z. Zhao. "Anode problems in pulsed power reactive sputtering of dielectrics." 42nd annual technical conference of the SVC, Chicago, 169 (1999).

[130] P. Frach, U. Heisig, C. Gottfried and H. Walde. "Aspects and results of long-term stable deposition of Al_2O_3 with high rate pulsed reactive magnetron sputtering." Surf. Coat. Technol 59, 177 (1993).

[131] S. Schiller, J. Goedicke, J. Reschke, V. Kirchhoff, S. Schneider and F. Milde. "Pulsed magnetron sputter technology." Surf. Coat. Technol 61, 331 (1993).

[132] A. Belkind, A. Freilich, J. Lopez, Z. Zhao, W. Zhu and K. Becker. "Characterization of pulsed dc magnetron sputtering plasmas." New J. Phys. 7, 90 (2005).

[133] P. Frach, K. Goedicke, C. Gottfried and H. Bartzsch. "A versatile coating tool for reactive in-line sputtering in different pulse modes." Surf. Coat. Technol 142, 628 (2001).

[134] Masahiko Naoe, Shun'ichi Yamanaka and Youichi Hoshi. "Facing targets type of sputtering method for deposition of magnetic metal films at low temperature and high rate." IEEE Trans. Magn. Vol. 16, Issue 5, 646 (1980)

[135] Nobuhiro Matsushita, Kenji Noma, Shigeki Nakagawa and Masahiko Naoe. "Plasma diagnosis and low-substrate-temperature deposition of Ba ferrite films in a damage-free sputtering apparatus with mixed gases." Vacuum, 51, 543 (1998)

[136] L. S. Hung, L. S. Liao, C. S. Lee, and S. T. Lee. "Sputter Deposition of Cathodes in Organic Light Emitting Diodes." J. Appl. Phys. 86, 4607 (1999).

[137] J. Kassabov, E. Atanassova, D. Dimitrov, and J. Vasileva. "Influence of helium plasma UV radiation on the channel mobility and charges in Si-SiO_2 structures." Thin Solid Films. 169, 43 (1989).

[138] A. M. Wróbel, and G. Czeremuszkin. "Effect of UV irradiation of the film during plasma chemical vapour deposition." Thin Solid Films. 216, 203 (1992).

[139] A. C. Fozza, J. Roch, J. E. Klemberg-Sapieha, A. Kruse, A. Holläder, and M.R. Wertheimer. "Oxidation and ablation of polymers by vacuum-UV radiation from low pressure plasmas." Nucl. Instrum. Methods. Phys. Res. B 131, 205 (1997).

[140] G. Gu, V. Bulović, P. E. Burrows and S. R Forrest. "Transparent organic light emitting devices." Appl. Phys. Lett. 68, 2606 (1996)

[141] V. Bulović, P. Tian, P. E. Burrows, M. R. Gokhale, S. R Forrest and M. E. Thompson. "A surface-emitting vacuum-deposited organic light emitting devices." Appl. Phys. Lett. 70, 2954 (1997)

[142] G. Gu, G. Parthasarathy, S. R Forrest. "A metal-free, full-color stacked organic light-emitting device." Appl. Phys. Lett. 74, 305 (1999)

[143] H. K. Kim, D. G. Kim, K. S. Lee, M. S. Huh, S. H. Jeong and K. I. Kim. "Plasma damage-free sputtering of indium tin oxide cathode layers for top-emitting organic light-emitting diodes." Appl. Phys. Lett. 86, 183503 (2005)

[144] H. Kanno, Y. Sun and S. R. Forrest. "High-efficiency top-emissive white-light-emitting organic eletrophosphorescent devices." Appl. Phys. Lett. 86, 263502-1 (2005)

[145] S. Y. Ryu, S. H. Choi, J. T. Kim, C. S. Kim, H. K. Baik and H. S. Jeong. "Highly efficient transparent organic light-emitting diodes by ion beam assisted deposition-prepared indium tin oxide cathode." Appl. Phys. Lett. 90, 033513 (2007)

[146] S. Mock. Projektarbeit, Rollex (2009)

[147] M. Pfeiffer. "Controlled doping of organic vacuum deposited dye layers: Basics and applications." PhD thesis, TU Dresden (1999)

[148] J. Blochwitz. "Organic light-emitting diodes with doped charge transport layers." Ph. D Thesis, TU Dresden (2001)

[149] R. Meerheim, S. Scholz, G. Schwartz, S. Reineke, S. Olthof, K. Walzer, and K. Leo, "Efficiency and lifetime enhancement of phosphorescent organic devices." Proc. SPIE 6999, 699917 (2008).

[150] Calculated using emissive thin film optics simulation (ETFOS) software, designed by CCP of Zurich University of Applied Sciences, Winterthur (ZHW).

[151] C. N. Li, C. Y. Kwong, A. B. Djurisic, P. T. Lai, P. C. Chui, W. K. Chan and S. Y. Liu. "Improved performance of OLEDs with ITO surface treatments." Thin solid films, 477, 57 (2005).

[152] K. P. Kim, A. M. Hussain, D. K. Hwang, S. H. Woo, H. K. Lyu, S. H. Baek, Y. M. Jang and J. H. Kim. "Work function modification of indium-tin oxide by surface plasma treatments using different gases." Jpn. J. Appl. Phys. 48, 021601 (2009).

[153] D. Lu, Y. Wu, J. Guo, G. Lu, Y. Wang and J. Shen. "Surface treatment of indium tin oxide by oxygen-plasma for organic light-emitting diodes." Mater. Sci. Eng., B97, 141 (2003)

[154] Z. Z. You and J. Y. Dong. "Effect of oxygen plasma treatment on the surface properties of tin-doped indium oxide substrates for polymer LEDs." J. Colloid Interface Sci. 300, 697 (2006).

[155] Edwin Hall. "On a new action of the magnet on electric current." Am. J. Math. 2, 287 (1879).

[156] L. J. Van der Pauw. "A method of measuring specific resistivity and Hall effect of discs of arbitrary shape." Philips Research Reports 13, 1 (1958).

[157] S. Scholz, C. Corten, K. Walzer, D. Kuckling and K. Leo. "Photochemical reactions in organic semiconductor thin films." Organic Electronics 8, 709 (2007).

[158] S. Scholz, R. Meerheim, B. Lüssem and K. Leo. "Laser desorption/ionization time-of-flight mass spectrometry: A predictive tool for the lifetime of organic light emitting devices." Appl. Phys. Lett. 94, 043314 (2009)

[159] S. Scholz, Q. Huang, M. Thomschke, S. Olthof, P. Sebastian, K. Walzer, K. Leo, s. Oswald, C. Corten and D. Kuckling. "Self-doping and partial oxidation of metal-on-organic interfaces for organic semiconductor devices studied by chemical analysis techniques." J. Appl. Phys. 104, 104502 (2008).

[160] E. S. Machlin. "Materials science in microelectronics, the relation between thin film processing and structure." Giro Press, Croton-on-Hudson, New York, p. 157 (1995)

[161] Y. Tomita. "Alternative transparent electrodes for organic light emitting diodes." Ph. D dissertation, Technische Universität Dresden. (2009).

[162] L. S. Hung, L. S. Liao, C. S. Lee, and S. T. Lee. "Sputter Deposition of Cathodes in Organic Light Emitting Diodes." J. Appl. Phys. 86, 4607 (1999).

[163] A. C. Dürr, F. Schreiber, M. Kelsch, H. D. Carstanjen and H. Dosch. "Morphology and thermal stability of metal contacts on crystalline organic thin films." Adv. Mater. 14, 961 (2002).

[164] R.Q. Zhang, C.S. Lee and S.T. Lee. "Effects of charging on electronic structure of the Alq3 molecule: the identification of carrier transport properties." Chemical Physics Letters 326, 413 (2000).

[165] H. Yasuda. "Plasma for modification of polymers." J. Macromol. Sci.-Chem. A10 (3) (1976) 383.

[166] F.M. d'Heurle. "Aluminum films deposited by rf sputtering." Metall. Trans. 1, 725 (1970)

[167] S. M. Rossnagel and K. L. Saenger. "Optical emission in magnetrons: Nonlinear aspects." J. Vac. Sci. Technol. A 7, 968 (1989)

[168] A. Belkind, Z. Zhao, D. Carter, L. Mahoney, G. McDonough, G. Roche, R. Scholl, and H. Walde. "Pulsed-DC reactive sputtering of dielectrics: Pulsing parameter effects." Society of vacuum coaters, 43rd annual technical conference proceedings, 86 (2000).

[169] R. Meerheim, K. Walzer, M. Pfeiffer and K. Leo. "Ultrastable and efficient red organic light emitting diodes with doped transport layers." Appl. Phys. Lett. 89, 061111 (2006).

[170] Q. Wang, C. Wang, Z. Wang, J. Zhang and D. He. "Stress relief patterns of hydrogenated amorphous carbon films grown by dc-pulse plasma chemical vapor deposition." Applied Surface Science 255, 1836 (2008)

[171] D. He, W. Cheng, J. Qin, J. Yue, E. Xie and G. Chen. "Relief of the internal stress in ternary boron carbonitride thin films." Applied Surface Science 191, 338 (2002)

[172] Z. J. Liu, N. Jiang, Y. G. Shen and X. Li. "Stress-induced surface damages in Ti-Si-N films grown by magnetron sputtering." Thin Solid Films 516, 7609 (2008)

[173] H. Sahm, C. Charton, R. Thielsch. "Oxidation behavior of thin silver films deposited on plastic web characterized by spectroscopic ellipsometry (SE)." Thin Solid Films 455–456, 819–823 (2004)

[174] N. Matsunami, Y. Yamamura, Y. Itikawa, N. Itoh, Y. Kazumata, S. Miyagawa, K. Morita, R. Shimizu and H Tawara. "Energy dependence of the ion-induced sputtering yields of monatomic solids." Atomic Data and Nuclear Data Tables 31, 1 (1984).

[175] R. Mientus and K. Ellmer. "Reactive magnetron sputtering of tin-doped indium oxide (ITO): Influence of argon pressure and plasma excitation mode." Surf. Coat. Technol. 142–144, 748 (2001).

[176] S. Honda, M. Watamori and K. Oura. "The effects of oxygen content on electrical and optical properties of indium tin oxide films fabricated by reactive sputtering." Thin Solid Films 281, 206 (1996).

[177] Z. Qiao and D. Mergel. "Comparison of radio-frequency and direct-current magnetron sputtered thin In2O3:Sn films." Thin Solid Films 484, 146 (2005).

[178] J. A. Dobrowolski, F. C. Ho, D. Menagh, R. Simpson and A. Waldorf. "Transparent, conducting indium tin oxide films formed on low or medium temperature substrates by ion-assisted deposition." Appl. Opt. 26, 5204 (1987).

[179] L. Meng, J. Gao, V. Teixeira and M. P. Santos. "Indium tin oxide thin films prepared by ion beam assisted deposition technique at different ion beam currents." Physica Status Solidi 205, 1961 (2008).

[180] P. F. Carcia, R. S. McLean, M. H. Reilly, Z. G. Li, L. J. Pillione and R. F. Messier. "Influence of energetic bombardment on stress, resistivity, and microstructure of indium tin oxide films grown by radio frequency magnetron sputtering on flexible polyester substrates." J. Vac. Sci. Technol. A 21, 745 (2003).

[181] T. C. Gorjanc, D. Leong, C. Py and D. Roth. "Room temperature deposition of ITO using r.f. magnetron sputtering." Thin Solid Films 413, 181 (2002).

[182] H. De Neve, J. Blondelle, R. Baets, P. Demeester, P. Van Daele, G. Borghs. "High efficiency planar microcavity LED's: Comparison of design and experiment." IEEE Photonics Technol. Letter. 7, 287 (1995).

[183] W. T. McKenna and L. Ward, "The optical properties of discontinuous silver films." Phys. Stat. Sol. (a) 68, K11 (1981).

[184] U. Kreibig and C. V. Fragstein. "The limitation of electron mean free path in small silver particles." Z. Physik 224, 307 (1969).

[185] J. C. Charton. "Schichtwachstum, optische und elektrische Eigenschaften von Silber auf PET-Folie." Doctor dissertation, Otto-von-Guericke Universität Magdeburg Germany (2002).

[186] K. Fehse. Projektarbeit. Hypoled (2008)

[187] A. Rogozin, N. Shevchenko, M. Vinnichenko, M. Seidel, A. Kolitsch and W. Möller. "Annealing of indium tin oxide films by electric current: Properties and structure evolution." Appl. Phys. Lett. 89, 061908 (2006).

[188] Q. Huang, K. Walzer, M. Pfeiffer, K. Leo, M. Hofmann and T. Stübinger. "Performance improvement of top-emitting organic light-emitting diodes by an organic capping layer: an experimental study", J. App. Phys. 100, 064507 (2006).

[189] M. Anwar, C. A. Hogarth and K. A. K. Lott. "ESR studies of MoO_3/In_2O_3 amorphous thin films deposited by co-evaporation." Phys. Stat. sol. (a) 113, 483 (1989).

[190] S. S. Sunu, E. Prabhu, V. Jayaraman, K. I. Gnanasekar, T. K. Seshagiri and T. Granasekaran. "Electrical conductivity and gas sensing properties of MoO_3." Sensors and Actuators B 101, 161 (2004).

[191] S. Kaleemulla, A. Sivasankar Reddy, S. Uthanna and P. Sreedhara Reddy. "Physical properties of flash evaporated In_2O_3 films prepared at different substrate temperatures." Materials Letters 61, 4309 (2007).

List of symbols

ε	Permittivity
σ	Electrical conductivity; collision cross section
λ	Wavelength; mean free path
ω	Plasma frequency
Y	Sputter yield
μ_e, μ_h	Electron mobility, hole mobility
η_{int}, η_{out}, η_{ext}	Internal, outcoupling, external quantum efficiency
a_0^*	Bohr radius
B	Magnetic field
E	Electric field
e	Elementary charge
F	Force
I	Current
J	Current density
L	Light output in lumens
m	Mass
n, p	Electron carrier concentration
N_t	Trap density of state
T	Temperature

List of abbreviations

AFM	Atomic force microscopy
AMOLED	Active matrix organic light emitting diode
BEOLED	Bottom emission organic light emitting diode
DC	Direct current
EBL	Exciton blocking layer
EDX	Energy dispersive X-ray spectroscopy
EML	Emission layer
ETL	Electron transport layer
FTS	Facing target sputtering
HOMO	Highest occupied molecular orbital
HTL	Hole transport layer
ITO	Indium tin oxide
LCD	Liquid crystal display
LUMO	Lowest unoccupied molecular orbital
OLED	Organic light emitting diode
OPV	Organic photovoltaic
OVPD	Organic vapor phase deposition
PLED	Polymer light emitting diode
PMOLED	Passive matrix organic light emitting diode
PVD	Physical vapor deposition
RF	Radio frequency
SCLC	Space charge limited current
SEM	Scanning electron microscopy
TCLC	Trap charge limited current
TCO	Transparent conducting oxide
TEOLED	Top emission organic light emitting diode
UPS	Ultraviolet photoelectron spectroscopy
UV	Ultraviolet
VUV	Vacuum ultraviolet
XPS	X-ray photoelectron spectroscopy
XRD	X-ray diffraction

Acknowledgement

First of all, I would like to express my gratitude to Prof. Hubert Lakner and Prof. Karl Leo for giving me a chance to be supervised by you. Your supervision of scientific problems and organization took a great part in my Ph. D work. The Doctor degree that I will be entitled by TU Dresden will be one of the most memorial things in my whole life.

I also want to appreciate all the members of the OMS group in the Fraunhofer Institute for Photonic Microsystems for support and collaboration. It was very precious time, which will not be forgettable in the rest of my life. I specially want to thank Dr. Christian May, Dr. Olaf Hild, Frank Löffler, Sebastian Franke, Dr. Michael Törker, and Dr. Ulrich Todt. Without you, I could not start, proceed, and finish this work. Your advices and helps made the main frame of this work, and Ihr habt auch Deutschland zu meiner zweiten warmen und fröhlichen Heimland gemacht. For the experiments, I thank

Dr. Karsten Fehse for the advices on OLED and cathode deposition,

Simone Giersch for the AFM and Ellipsometer measurement,

Dr. Bolko Schöneich for the evaluation of the samples with Ellipsometrie,

Thomas Schmitt for the OLED fabrication,

Claudia Lorenz for the *I-V-L* measurement,

Rüdiger Hermann and Gunter Harzendorf for the lifetime measurement,

Alexander Schmid for the data collection,

Dr. Wolfram Pufe for the SEM/EDX measurement,

Dr. Jan Uwe Schmidt for the XRD measurement,

Dr. Sebastian Scholz for the LDI-TOF-MS measurements,

John Falteich for the discussion on sputtering and the DMS setup,

the project partners Applied materials GmbH and OSRAM GmbH,

and the people who were directly and indirectly related to this work.

All these work could, however, be done because of my parent. They gave a birth to me, eyes to see this beautiful world, a nose to smell Schweinshaxe being roasted, a mouth to eat Kimchi of my mother, a body to run ahead, and a strong spirit to maintain me. I regard them with respect and attribute the entire honour to them.

I also specially thank to Hee Young who stayed by me in a far distance during my Ph. D course. I wish you a lot of success for your Ph. D work.

Versicherung

Hiermit versichere ich, daß ich die vorliegende Arbeit ohne unzulässige Hilfe Dritter und ohne Benutzung anderer als der angegebenen Hilfsmittel angefertigt habe; die aus fremden Quellen direkt oder indirekt übernommenen Gedanken sind als solche kenntlich gemacht.

Die Arbeit wurde bisher weder im Inland noch im Ausland in gleicher oder ähnlicher Form einer anderen Prüfungsbehörde vorgelegt.

Diese Arbeit wurde am Fraunhofer Institut für Photonische Mikrosysteme unter der wissenschaftlichen Betreuung von Prof. Dr. Hubert Lakner und Karl Leo angefertigt.

Ich erkenne die Promotionsordnung der Fakultät Elektrotechnik der Technischen Universität Dresden vom 1. April 2006 an.

Dresden, am 28. 6. 2011

Tae Hyun Gil